Textbook 2B

Maths — No Problem!

English National Curriculum

Consultant Editors
Dr Wong Khoon Yoong
Dr Julie Alderton

Series Editors
Adam Gifford
Andy Psarianos

Authors
Adam Gifford
Gemma Meharg

MATHS NO PROBLEM!

Published by Maths — No Problem!

Printed in the United Kingdom
First Printing, 2022

ISBN 978-1-913458-52-2

Maths — No Problem!
hello@mathsnoproblem.com
www.mathsnoproblem.com

Acknowledgements
The publisher would like to thank Dr Anne Hermanson, Alex Laurie and Judy Hornigold for their contributions.

Preface

Welcome to Maths — No Problem!

Maths — No Problem! is a comprehensive series of textbooks and workbooks supported by detailed online teacher guidance to deliver maths mastery in primary schools. The materials use a spiral design with carefully built-up mathematical concepts and processes adapted from the maths mastery approaches used in Singapore. An integral part of the learning process is the Concrete, Pictorial, Abstract (CPA) approach, which forms the backbone of the materials developed for this series.

Maths — No Problem! incorporates the use of concrete aids and manipulatives, problem solving, group work, individual practice and journaling.

Meet our Maths — No Problem! friends who will join you on your journey through this textbook.

The Maths — No Problem! approach is exemplified throughout the chapters by the following teaching elements:

Chapter Opener

The chapter opener depicts familiar events or contexts that serve as an introduction for pupils to the concept covered in the chapter. It can also be used as an additional prompt for journal writing or an additional exercise for pupils.

Explore

Explore is an introductory activity for pupils, which includes questions related to various lesson objectives.

Master

Each lesson contains a Master section that develops lesson objectives by introducing new concepts through a CPA approach using manipulatives and engaging pictures. Guided examples are provided to reinforce the learning objective.

Activity Time

Activity Time provides pupils with opportunities to work as individuals or in small groups to explore mathematical concepts through games and investigations.

Guided Practice

Guided Practice contains questions for further consolidation and the immediate evaluation of pupils' learning.

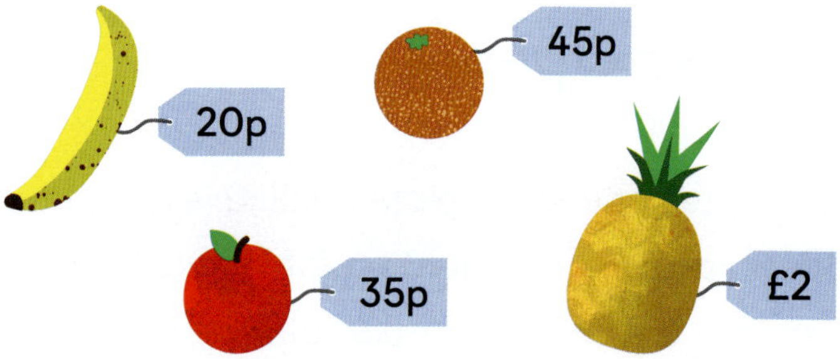

Mind Challenge

Mind Challenges are non-routine questions for pupils to apply relevant problem-solving skills and develop higher-order thinking skills.

Ravi

Elliott

My Maths Journal

My Maths Journal provides pupils with opportunities to show their understanding of the mathematical concepts learnt and to record their learning.

Review

Review allows pupils to reflect on their own learning after each chapter, with support from the teacher.

Review

I can...

- ☐ name coins and notes.
- ☐ count to find an amount of money.
- ☐ show amounts of money in different ways.
- ☐ exchange coins and notes.
- ☐ compare different amounts of money.
- ☐ calculate change.
- ☐ solve word problems involving money.

Contents

Chapter 11 2D Shapes

Chapter 12 3D Shapes

Chapter 13

Fractions

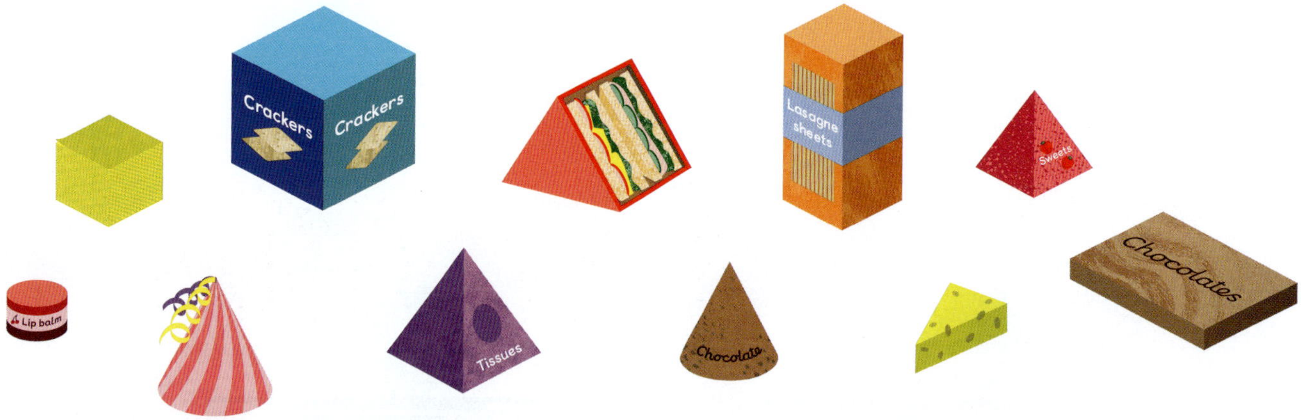

Chapter 14 Time

Chapter 15 Volume

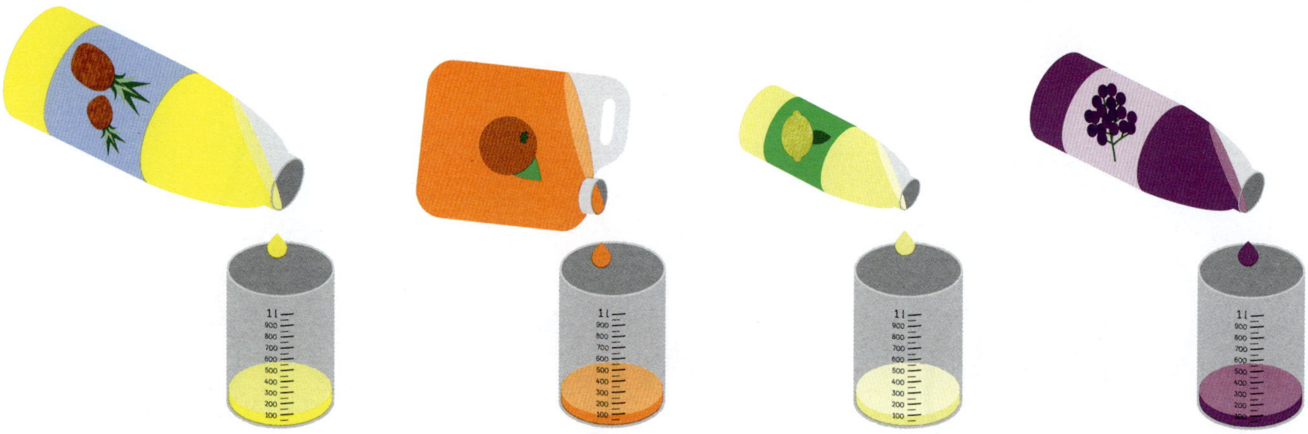

How can we find out how many strawberries are left?

Chapter 9
More Word Problems

Solving Word Problems

Lesson 1

Explore

How can we find out how many strawberries are left?

Master

1 (a) Find the total number of strawberries Hannah and Sam ate.

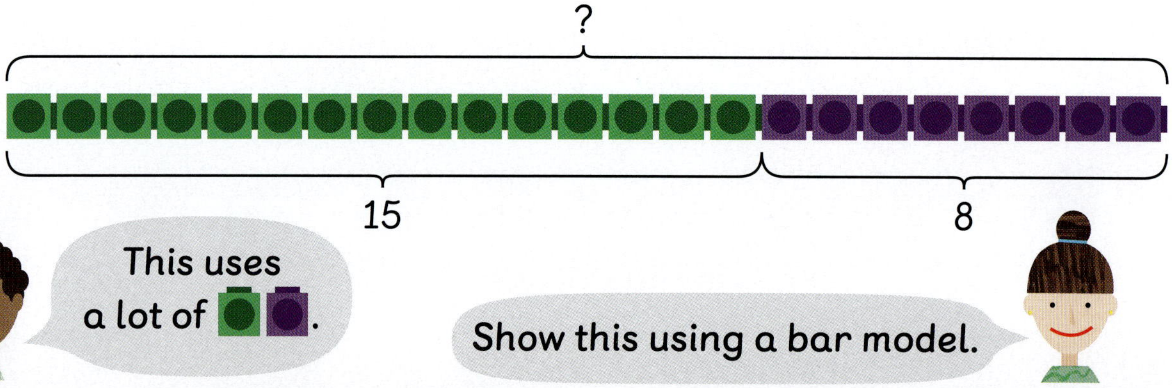

This uses a lot of .

Show this using a bar model.

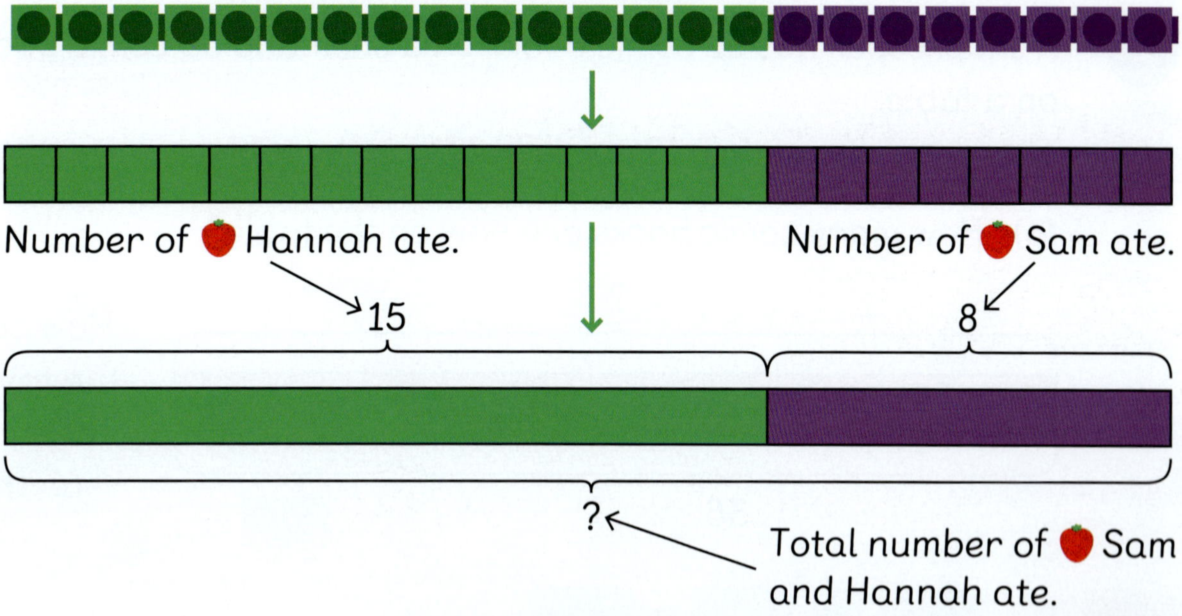

Add to find the total number of strawberries Hannah and Sam ate.

15 + 8 = 23

Hannah and Sam ate 23 strawberries.

(b) Find the number of strawberries left.

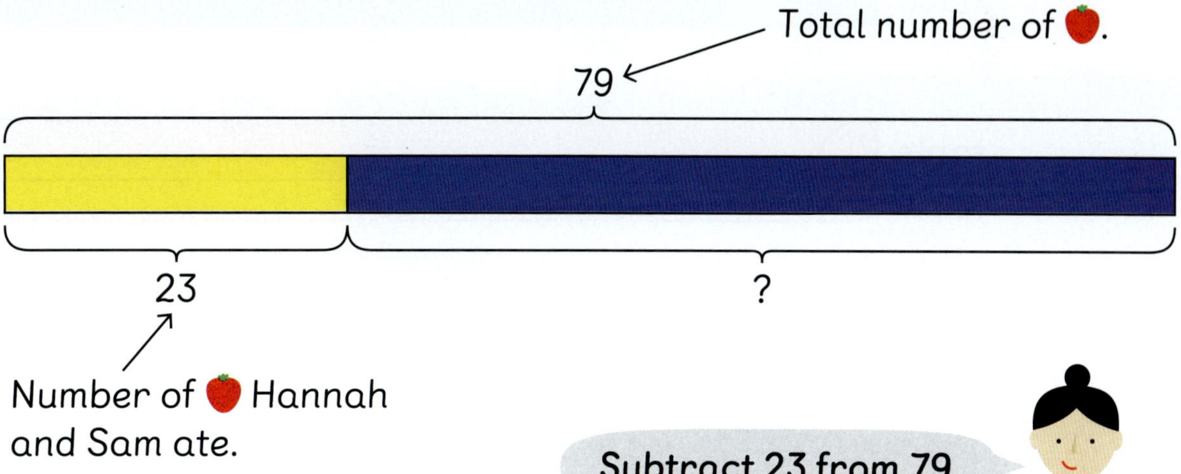

Subtract 23 from 79.

79 − 23 = 56

There are 56 strawberries left.

2 A shopkeeper has 52 comic books on a shelf and 36 comic books on a table.
Then he puts another 12 comic books on the table.

(a) How many comic books are now on the table?

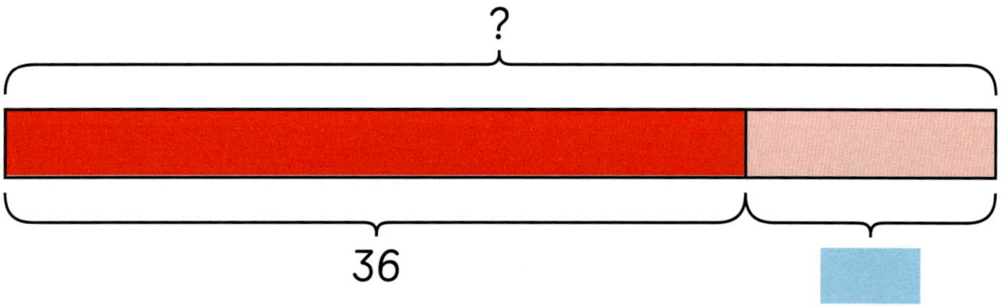

___ ◯ ___ = ___

There are ___ comic books on the table.

(b) How many more comic books are on the shelf than on the table?

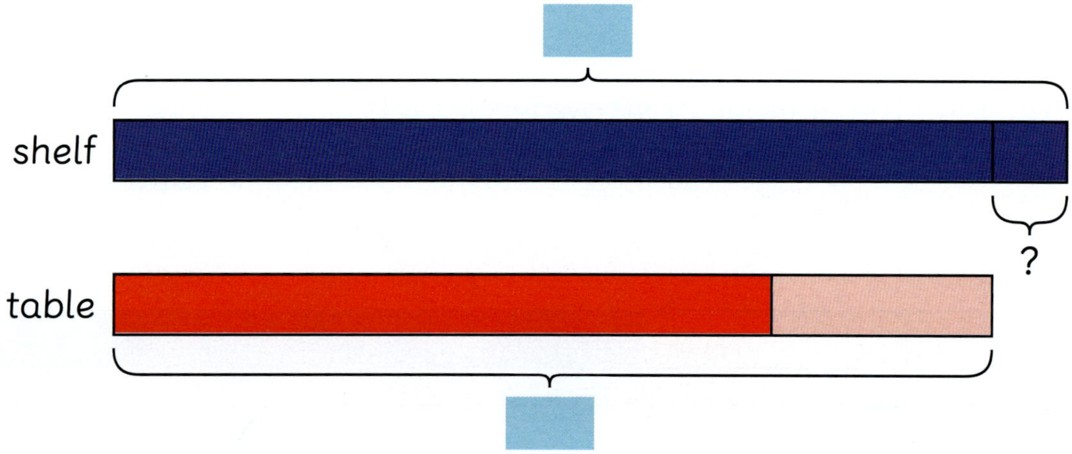

___ ◯ ___ = ___

There are ___ more comic books on the shelf than on the table.

Guided Practice

Solve.

1 There are 32 outdoor plants and 35 indoor plants for sale at a market.

(a) How many plants are for sale at the market altogether?

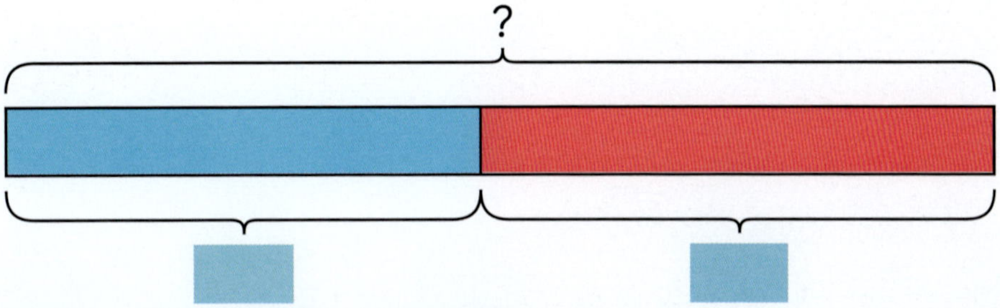

____ ◯ ____ = ____

There are ____ plants for sale at the market altogether.

(b) 56 of the plants are sold at the market.
How many plants are left?

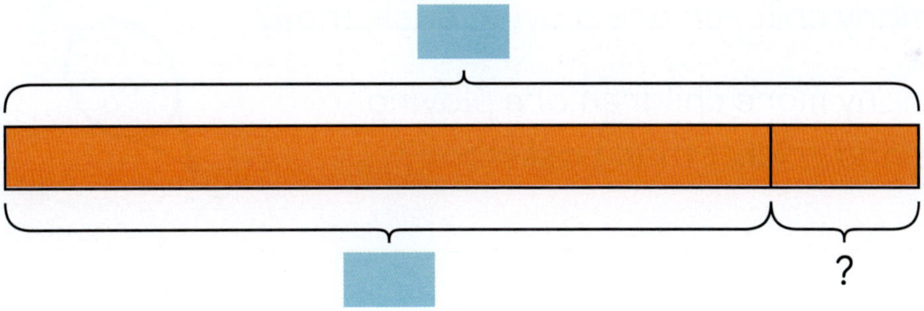

____ ◯ ____ = ____

There are ____ plants left.

2 There are 67 children on a field playing football or rounders.
There are 42 children playing rounders.

(a) How many children are playing football?

(b) How many more children are playing rounders than football?

3 There are 87 children playing sport at school.
34 children are playing tennis and 25 children are playing cricket.
The rest of the children are playing basketball.

(a) How many children are playing basketball?

(b) How many more children are playing tennis and cricket than basketball?

Complete Worksheet **1**: pages **1–4**

Solving Word Problems

Lesson 2

Explore

On Monday, Ravi swam 35 m.
On Tuesday, he swam 23 m further than he swam on Monday.
What distance did Ravi swim on Tuesday?

Master

1 (a) Find the distance Ravi swam on Tuesday.

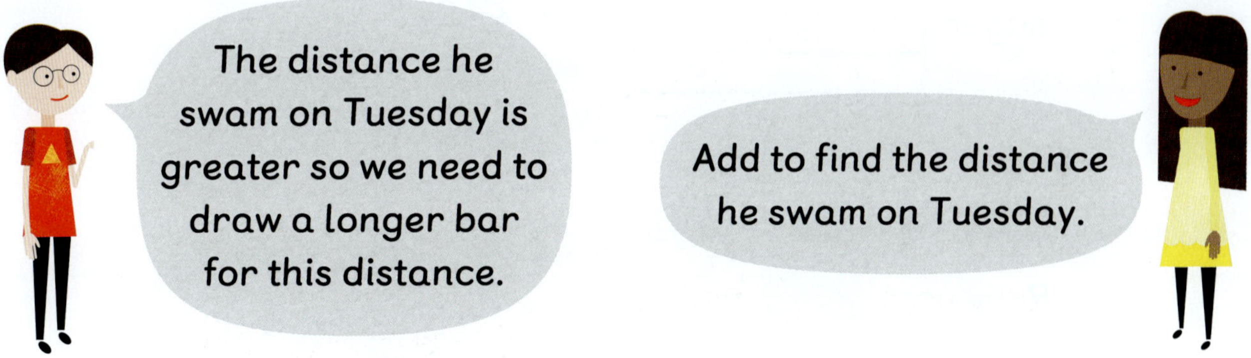

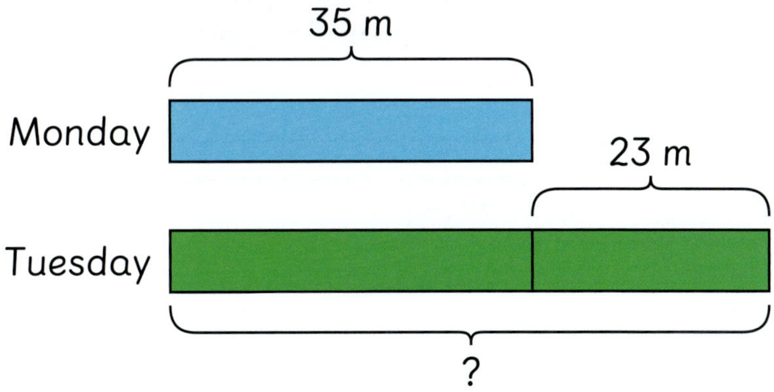

35 + 23 = 58
On Tuesday, Ravi swam 58 m.

(b) What is the total distance Ravi swam on Monday and Tuesday?

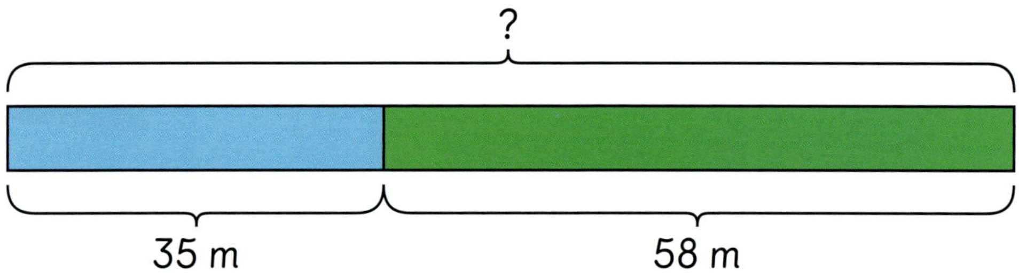

35 + 58 = 93

The total distance Ravi swam on Monday and Tuesday is 93 m.

2 Ruby makes scones using 64 g of butter.
She uses 29 g less sugar than butter.

(a) What is the mass of the sugar she uses?

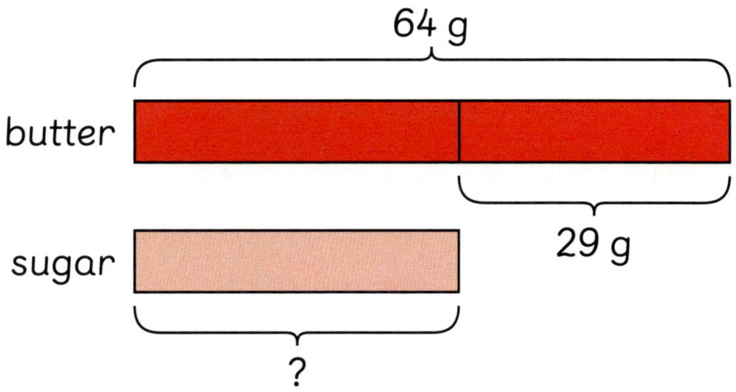

☐ ◯ ☐ = ☐

Ruby uses ☐ g of sugar.

(b) What is the total mass of the butter and sugar Ruby uses?

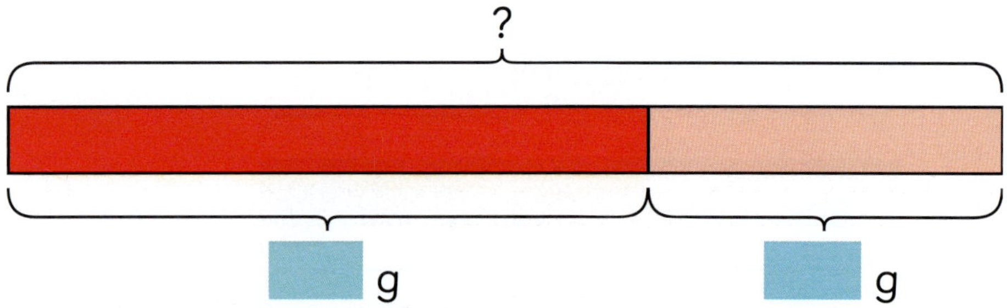

☐ ◯ ☐ = ☐

The total mass of the butter and sugar Ruby uses is ☐ g.

Guided Practice

Solve.

1 A box of books is 38 cm tall. The box of books is 19 cm taller than a box of rulers.

(a) What is the height of the box of rulers?

(b) The box of rulers is put on top of the box of books.

What is the total height of the two boxes?

2 A bakery has a 50 kg bag of flour.
It uses 23 kg of the flour on Saturday.
It uses 12 kg more flour on Saturday than on Sunday.

(a) What is the mass of the flour the bakery uses on Sunday?

(b) What is the mass of the flour left in the bag?

3 The distance between the library and the shop is 21 m less than the distance between Jacob's house and the library.

What is the distance between Jacob's house and the shop?

Complete Worksheet **2**: pages **5–8**

Solving Word Problems

Lesson 3

Explore

Hannah carried out a traffic survey. In one hour, she counted 46 cars and 12 fewer motorbikes than cars. How many vehicles did Hannah count in total?

Master

1

Start by finding the number of motorbikes.

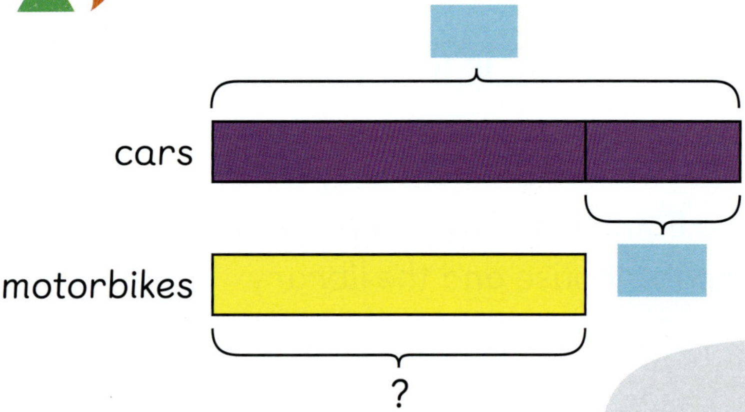

Do we add or subtract to find the number of motorbikes?

☐ ○ ☐ = ☐

Hannah counted ☐ motorbikes.

Then, find the total number of vehicles.

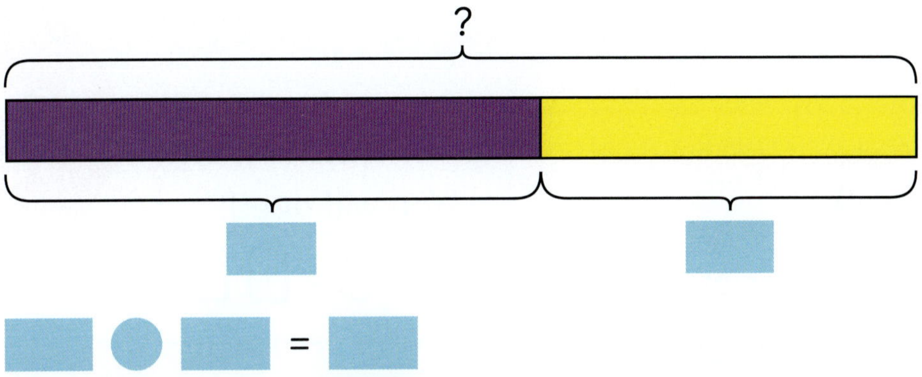

Hannah counted ___ vehicles in total.

2 There are 24 more children in a netball group than in a tennis group.
There are 48 children in the netball group.
What is the total number of children in both groups?

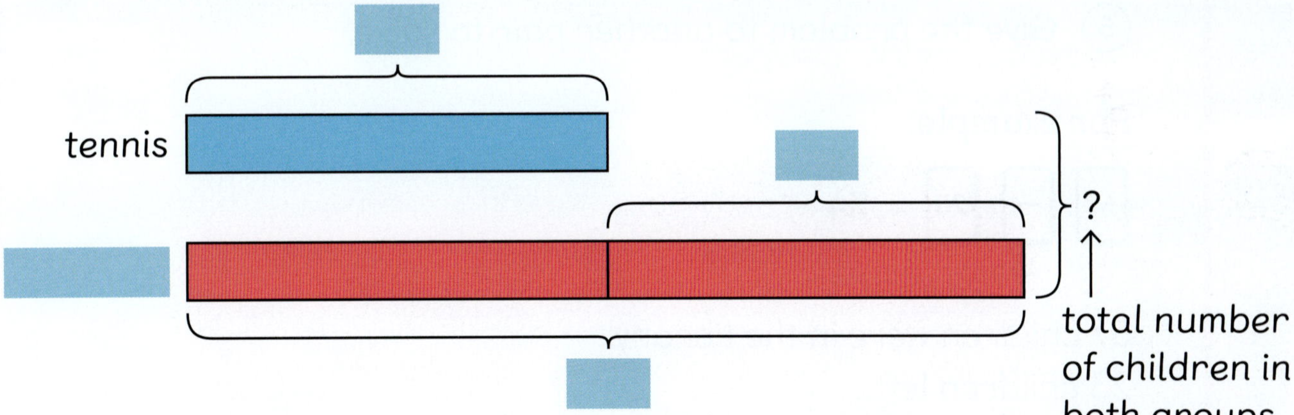

(a) How many children are in the tennis group?

There are ___ children in the tennis group.

(b) How many children are in both groups?

There are ___ children in both groups.

Activity Time

Partner work

You will need: 20 40 cards, 50 70 cards, + card, − card

1. Make three piles of cards: 20 40, 50 70, + −.
2. Turn over one card from each pile to make an addition or subtraction equation.
3. Solve the equation.
4. Write a word problem to match the equation.
5. Give the problem to another pair to solve.

For example:

57 − 23 = 34

57 children were in the library.
23 children left.
How many children were left in the library?

There were 34 children left in the library.

Guided Practice

Solve.

1 Hannah swam 28 lengths of a pool during the weekend. She swam 4 more lengths than Lulu swam during the weekend.

(a) How many lengths did Lulu swim during the weekend?

(b) How many lengths did Hannah and Lulu swim altogether during the weekend?

2 There are 56 children in the school hall.
There are 16 more children in the library than the school hall.
How many children are in the school hall and library altogether?

3 There are 75 crayons on a table.
There are 13 fewer crayons on the table than in a tray.
How many crayons are on the table and in the tray in total?

Complete Worksheet **3**: pages **9–10**

Solving Word Problems

Lesson 4

Explore

An ice cream seller sold 32 ice cream cones over one weekend.
She sold 6 more ice cream cones on Sunday than on Saturday.
How many ice cream cones did she sell on each day?

Master

1 Method 1

Saturday	Sunday	Total
10	16	10 + 16 = 26
11	17	11 + 17 = 28
12	18	12 + 18 = 30
13	19	13 + 19 = 32

She sold 13 ice cream cones on Saturday.
She sold 19 ice cream cones on Sunday.

We can guess how many were sold on each day.

We know there were 6 more sold on Sunday.

Then, we can check if the total number sold was 32.

Method 2

Show the problem using a bar model.

We know the difference is 6.

Find two numbers that are the same and have a sum of 26.

11 + 11 = 22
12 + 12 = 24
13 + 13 = 26

Add ▭ and 13 to find the total number sold on Sunday.

She sold 13 ice cream cones on Saturday.
She sold 19 ice cream cones on Sunday.

2 A bus has some passengers on it.
10 passengers get on at the Park Street stop.
At the next stop, 15 passengers get off.
There are now 40 passengers on the bus.
How many passengers were on the bus before it stopped at Park Street?

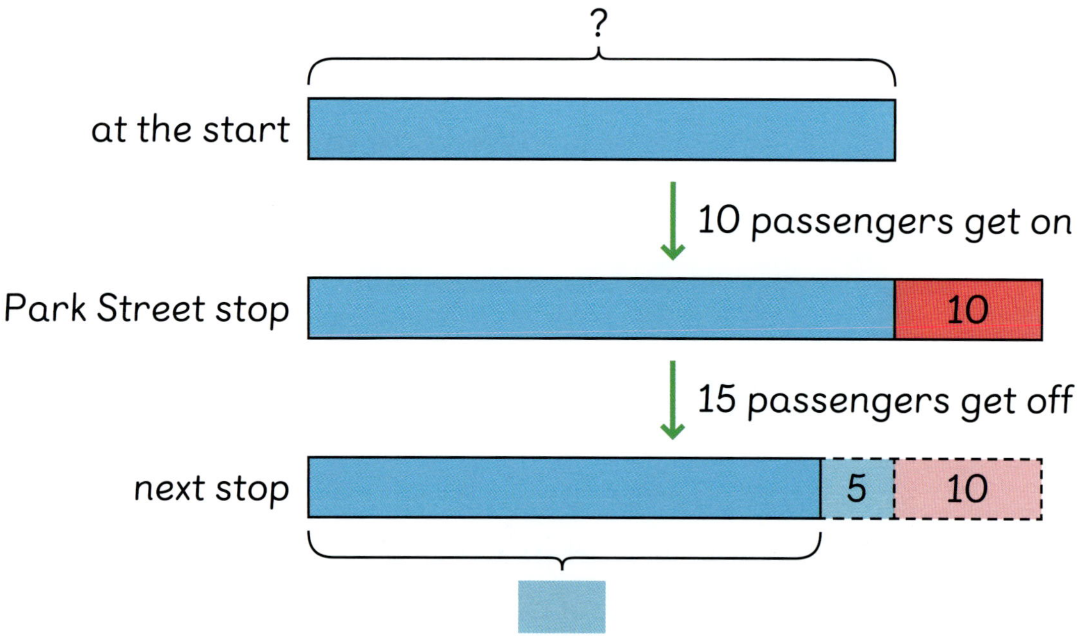

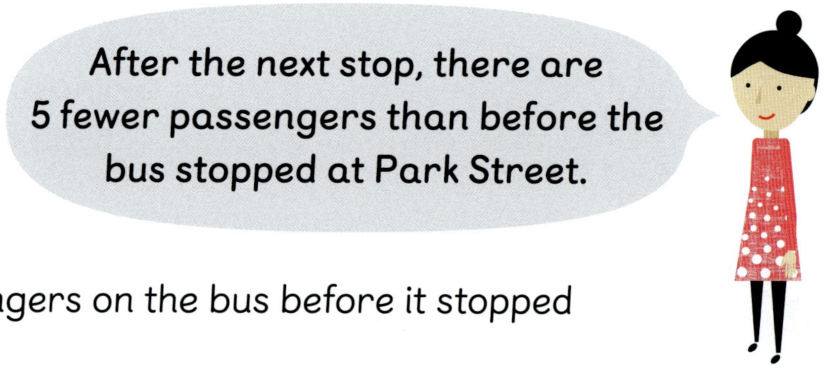

There were ____ passengers on the bus before it stopped at Park Street.

Guided Practice

Solve.

1. Elliott's backpack is 4 kg lighter than his suitcase.
What is the mass of his backpack?

2.

What is the mass of the  ?

3. Mr Nightingale had some ribbon. He bought 5 m more ribbon.
He then used 20 m of the ribbon to wrap prizes for the school fair.
He had 3 m of ribbon left at the end.
What length of ribbon did he have to start with?

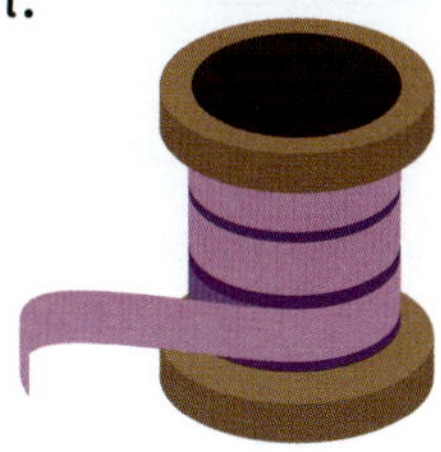

Complete Worksheet **4**: pages **11–13**

Mind Challenge

Find the mass of  .

Find the mass of  .

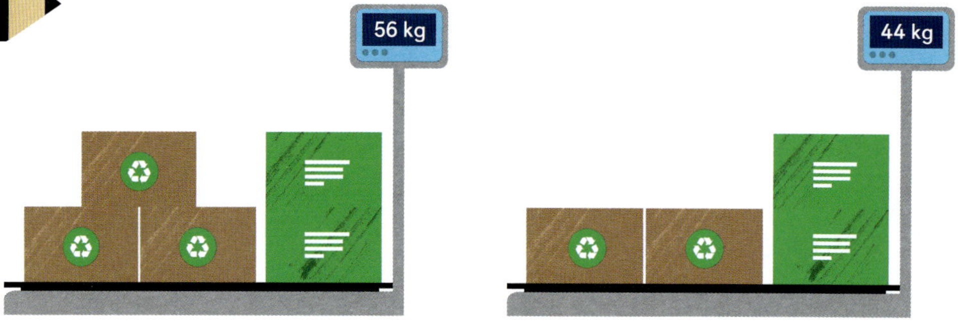

My Maths Journal

A sheep and a dog have a total mass of 65 kg.
The sheep is 35 kg heavier than the dog.
How heavy is each animal?

Guess and Check

Sheep	Dog	Total
		+ =
		+ =
		+ =
		+ =

Bar Model

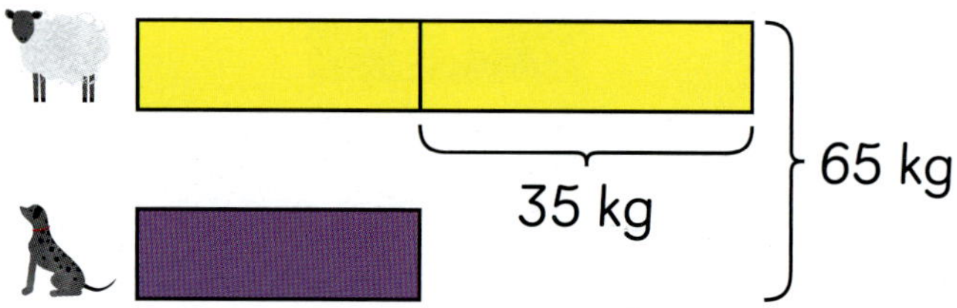

Which method would you choose to solve the problem?
Why would you choose that method?

Review

I can...

- ☐ draw models for different situations.
- ☐ solve word problems.
- ☐ use a guess-and-check strategy to solve problems.

Who has saved the most money?

Chapter 10
Money

Writing Amounts of Money

Lesson 1

Explore

Name the coins and notes each child has.

What does £ mean?
What does p mean?

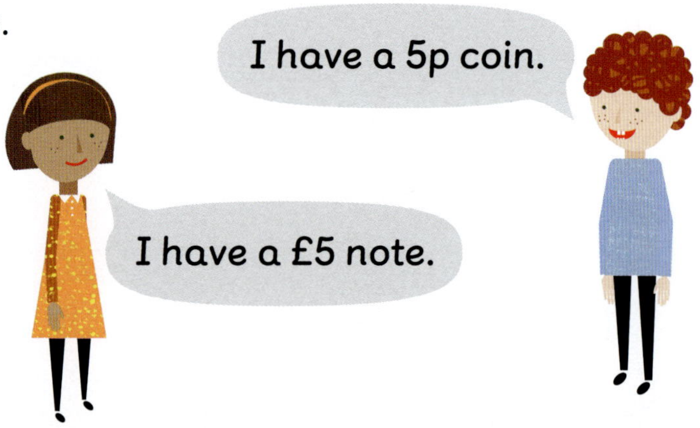

Master

1 has these coins.

This is a one penny coin.
We write it as **1p**.

p is the symbol for penny and pence.

2p
two pence

5p
five pence

10p
ten pence

Penny is the singular form of pence.

2 has these coins.

This is a one pound coin.
We write it as **£1**.

£ is the symbol for pound.

20p
twenty pence

50p
fifty pence

£2
two pounds

3  has these notes.

£5
five pounds

£10
ten pounds

£20
twenty pounds

£50
fifty pounds

p stands for penny and **pence**.
£ stands for **pounds**.

Activity Time

Partner work

You will need:

1. Place the cards face down in a pile.
2. Roll the dice or pick a card.
3. Show the correct amount of money to match the amount on the dice or card.
   Say the amount.
4. Ask your partner to check you are correct.
5. Take turns.

Guided Practice

1 Write the amount of money in words using **pounds** or **pence**.

(a)

(b)

(c)

(d)

(e)

(f)

2 Write the amount of money in numbers using **£** or **p**.

(a)

(b)

(c)

(d)

(e)

(f)

Complete Worksheet **1**: pages **17–18**

Counting Money Using Notes

Lesson 2

Explore

Two classes collected money for the school charity fund.
Which class collected more money?

Master

1

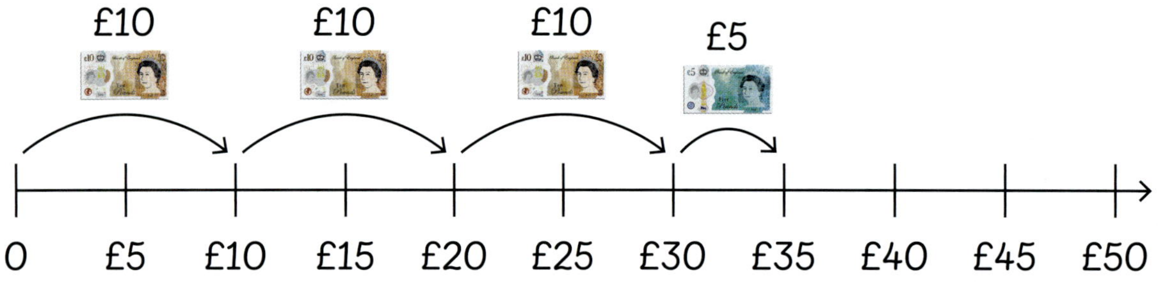

Class 5 collected £35.

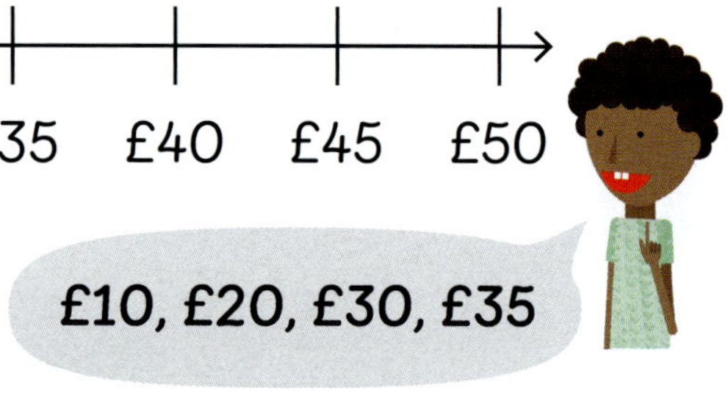

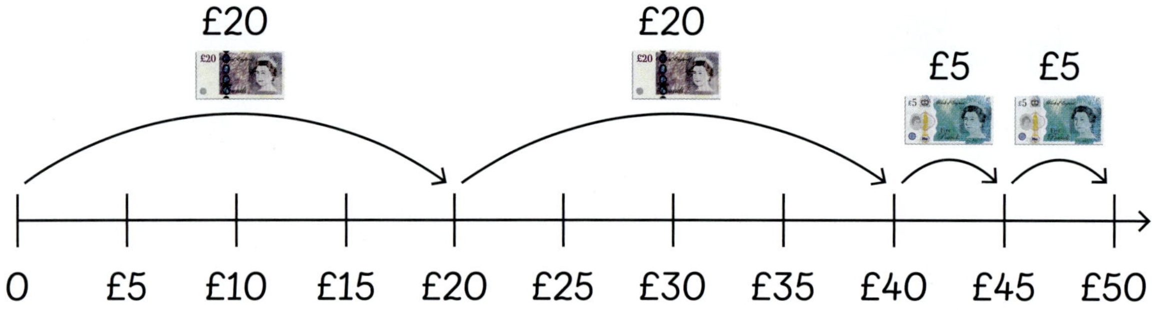

Class 6 collected £50.

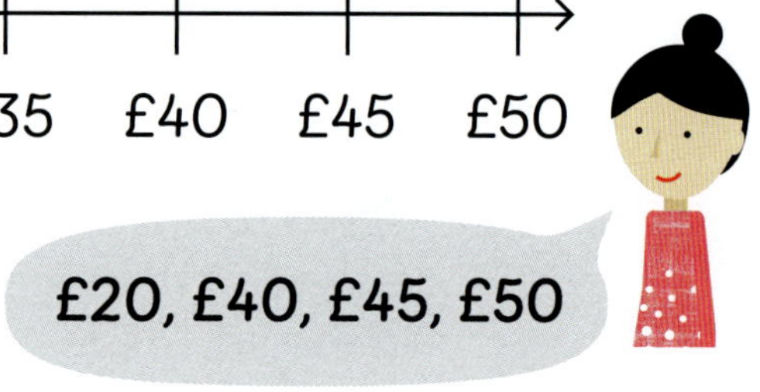

Class 6 collected more money than Class 5.

2 How much money did Class 3 and Class 4 each collect?

Class 3

Class 4

Class 3 collected £ ______.

Class 4 collected £ ______.

Has Class 4 collected a greater amount of money?

Class 3 and Class 4 collected an equal amount of money.

Guided Practice

1 What amount of money is shown?

(a)  ______

(b)  ______

2 (a) Show £25.

(b) Show £40.

(c) Show £50.

Did you and your partner show each amount of money in the same way? Who used more notes?

Complete Worksheet **2**: pages **19–20**

Counting Money Using Coins

Lesson 3

Explore

Holly and Oak each bought a toy using these coins.

Holly

Oak

How much did each child spend?

Master

1

£1 → £1 and 50p → £1 and 60p → £1 and 70p → £1 and 75p

Holly spent £1 and 75p.

Oak used fewer coins. Did she spend less money?

£2 → £2 and 20p → £2 and 30p → £2 and 35p

Oak spent £2 and 35p.

2 How much did Jacob spend on each sticker?

Jacob spent ______ on  .

Jacob spent ______ on  .

Guided Practice

1 Count and write how much money is shown.

(a) 

______

(b) 

______

(c) 

______

(d) 

______

2 (a) Show 85p.

(b) Show 64p.

(c) Show 37p.

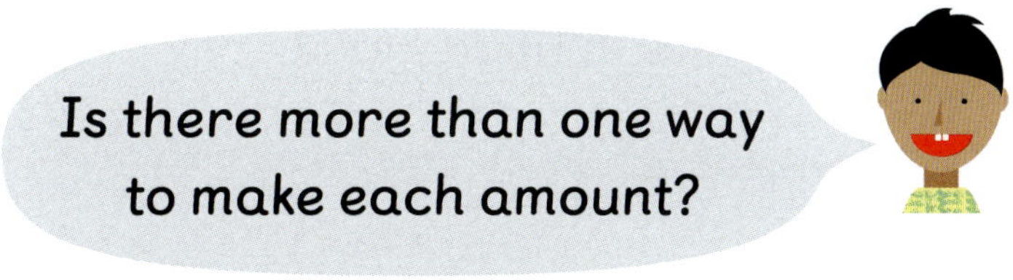

Complete Worksheet **3**: pages **21–22**

Counting Money

Lesson 4

Explore

Elliott wants to buy a cinema ticket.

Does he have enough money?

Master

1 Count on to find the amount of money Elliott has.

£5   £6   £6 and 50p   £6 and 70p   £6 and 80p   £6 and 85p

Elliott has £6 and 85p.
He does not have enough money.

£6 and 85p is less than £6 and 95p.

2 Can Elliott buy a cinema ticket with the following amount of money?

Start by counting the pounds.

Then, count the pence.

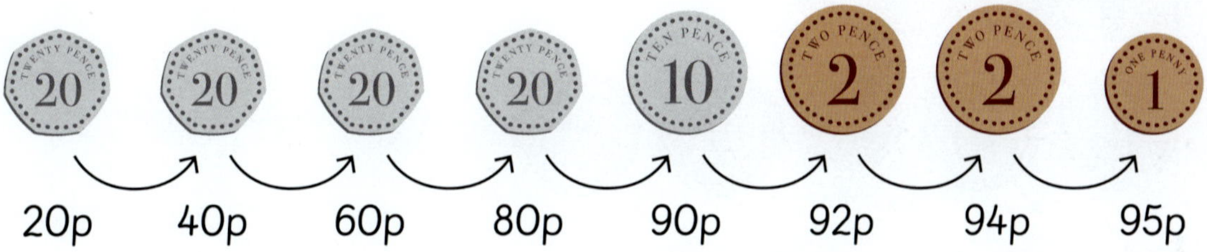

The amount is £6 and 95p.
Elliott can buy a cinema ticket.

Activity Time

Group work (4 pupils)

You will need:

1. One pupil acts as the shopkeeper.
2. The other pupils take turns to buy an item from the shop.
3. The pupil buying the item counts the correct amount of money and pays for it.
4. The shopkeeper checks the amount of money is correct.
5. Pupils take turns to act as the shopkeeper.

Guided Practice

1 Count and write how much money is shown.

(a) 

(b) 

(c) 

2 (a) Show £6 and 40p.

(b) Show £94 and 65p.

Complete Worksheet **4**: pages **23–24**

Showing Equal Amounts of Money

Lesson 5

Explore

Hannah's purse

Lulu's purse

Who has enough money to buy the sandwich?

Master

1 Count the amount of money Hannah has.

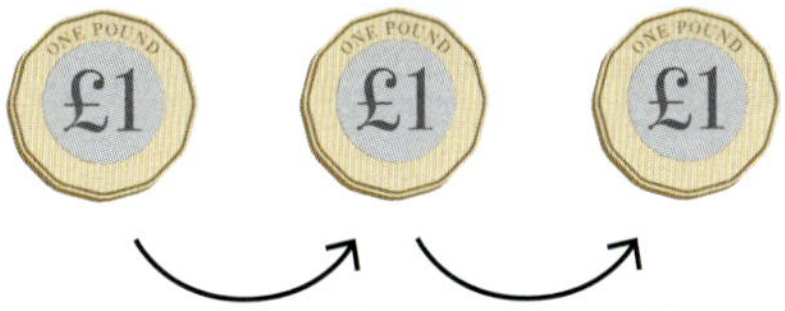

£1 £2 £3

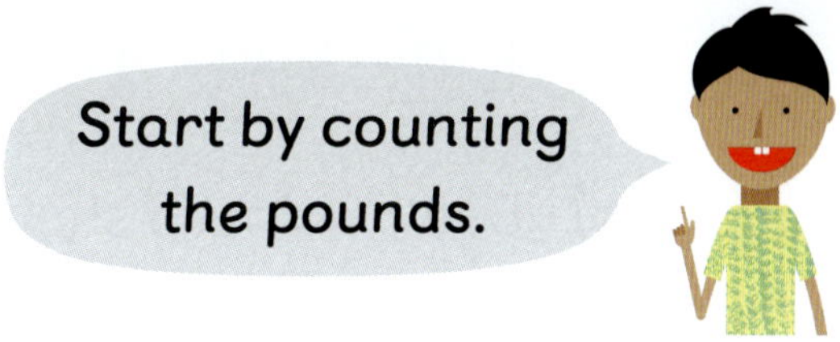

50p 60p 65p 67p 69p 70p

Hannah has £3 and 70p.

2 Count the amount of money Lulu has.

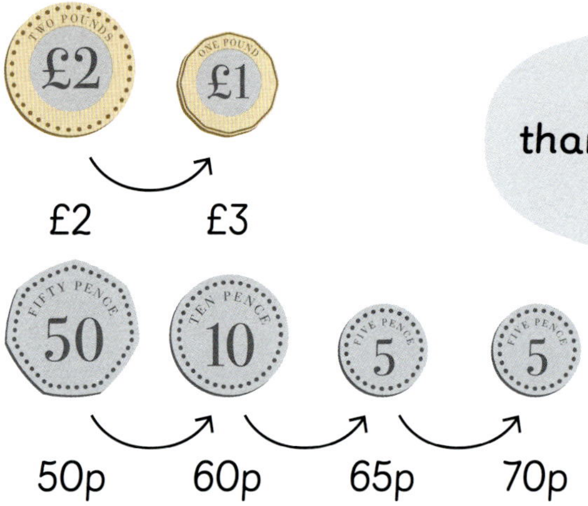

Lulu has £3 and 70p.

Hannah and Lulu each have £3 and 70p.
They have an equal amount of money.
Both of them have enough money to buy the sandwich.

Guided Practice

1 Show each amount of money in 2 different ways.

(a) £2 (b) £10

2 Show each amount of money using a different number of coins.

(a) Use 4 coins to show the following amount.

(b) Use 5 coins to show the following amount.

(c) Use 6 coins to show the following amount.

Complete Worksheet **5**: pages **25–26**

Exchanging Money

Lesson 6

Explore

Lulu has a  .

She wants to give her friends some money and keep some for herself.

How can she exchange the  for coins of the same total value?

Master

1 

Which coins are equal to £2?

£2 is equal to 2 one-pound coins.

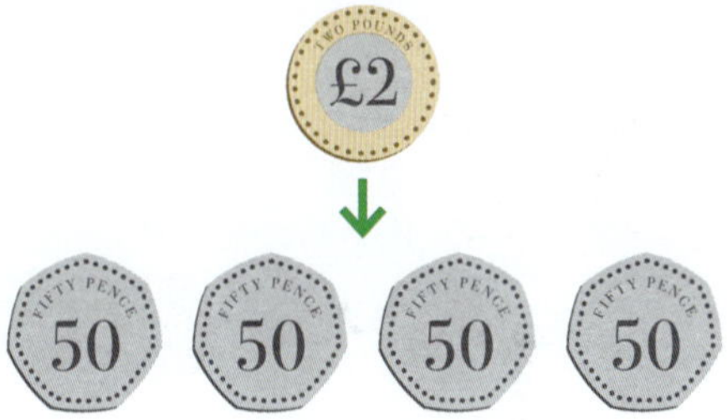

£2 is equal to 4 fifty-pence coins.

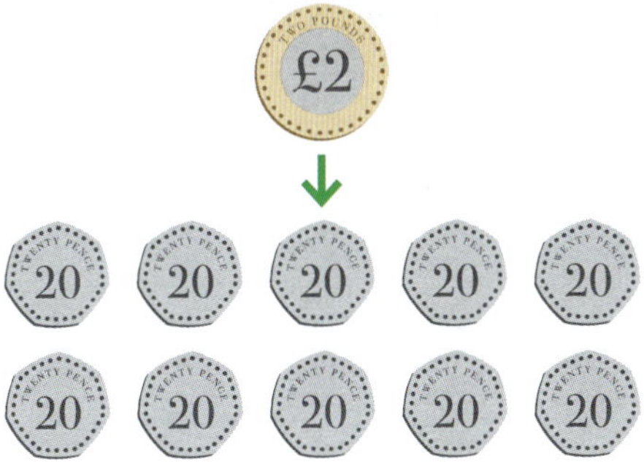

£2 is equal to 10 twenty-pence coins.

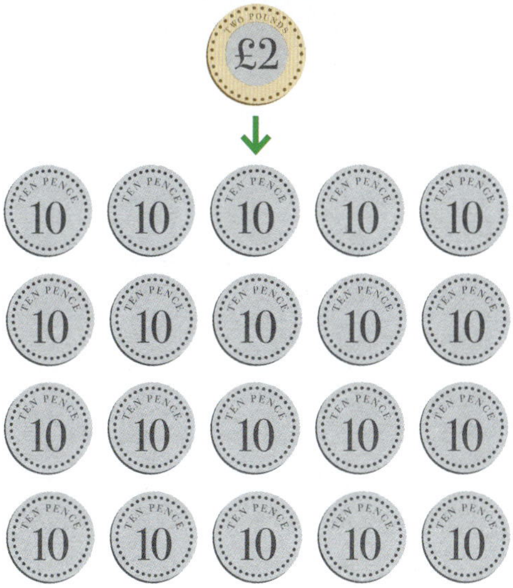

£2 is equal to 20 ten-pence coins.

2

We can exchange a coin for other coins that have the same value.

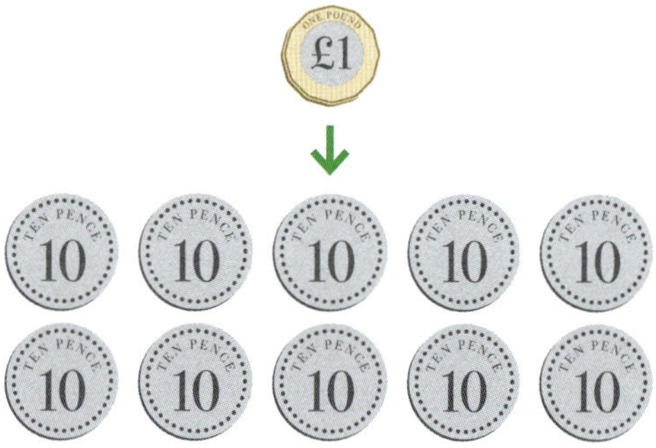

£1 is equal to 10 ten-pence coins.

50p is equal to 5 ten-pence coins.

20p is equal to 2 ten-pence coins.

10p is equal to 2 five-pence coins.

5p is equal to 5 one-penny coins.

2p is equal to 2 one-penny coins.

Guided Practice

1 Show the value of each amount in 3 different ways using coins.

(a) 

(b) 

(c) 

(d) 

2 Match.

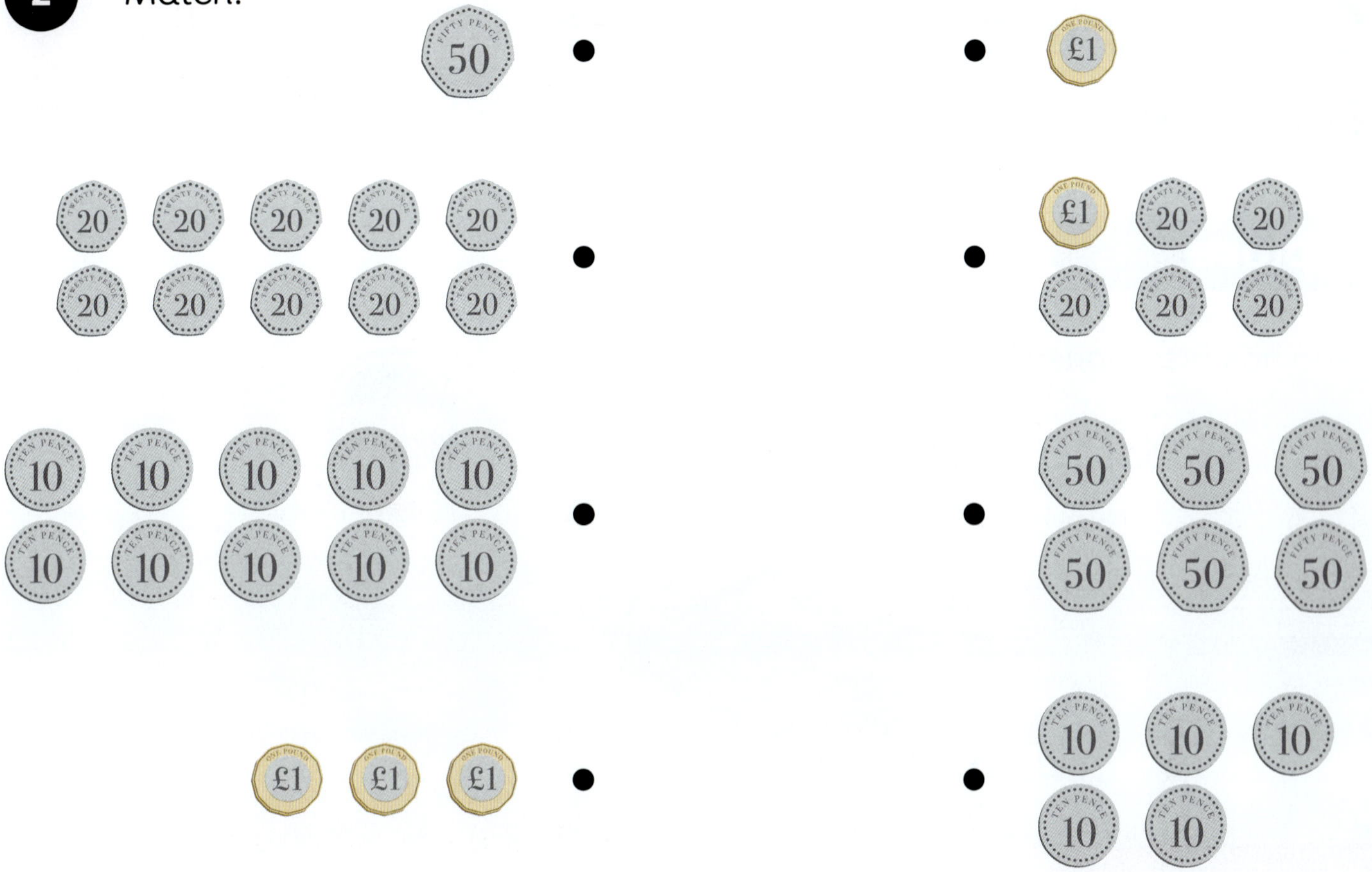

3 Ravi has some money in his pocket.
Show all the possible ways Ravi could have 10p in his pocket.

4 Oak exchanges £1 for 5 coins that have the same total value.

Find 3 different ways a £1 can be exchanged.

Complete Worksheet 6: pages 27–28

Comparing Amounts of Money

Lesson 7

Explore

Who has less money?

Master

1

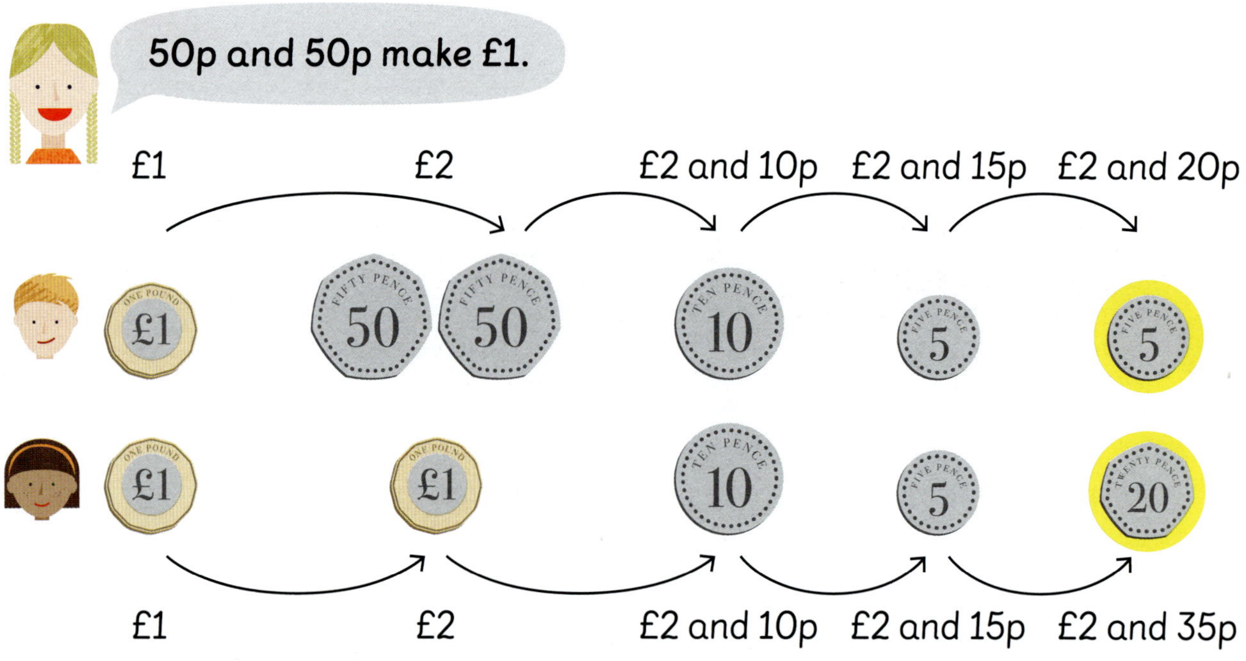

Sam has less money than Amira.

2 Compare the prices of the ice cream cones.

£2 is more than £1.
£2 and 15p is more than £1 and 75p.

3 Count and compare the amounts of money in the purses.

£1 and ___ p   £ ___ and 10p

£1 and ___ p is less than £ ___ and 10p.

£ ___ and 10p is more than £1 and ___ p.

Guided Practice

Compare.

Plant A

Plant B

Plant C

1. Which costs more, Plant B or Plant C?

2. Which costs more, Plant C or Plant A?

3. Which costs less, Plant A or Plant B?

4. Arrange the plants in price order.
   Start with the plant that costs the least.

   ____ , ____ , ____

Complete Worksheet **7**: pages **29–30**

Calculating Total Amount

Lesson 8

Explore

Charles and his dad are having lunch at a café.

MAINS		SAUCES	
Spicy noodles	£8	Sweet chilli	30p
Spicy rice	£9	Chilli sauce	35p
Spicy soup	£6	Soy sauce	20p

Charles chooses the spicy noodles and his dad chooses the spicy rice.
What is the total cost of their main meals?

Master

1 **Method 1** Count.

spicy noodles

spicy rice

£10 + £7 = £17
The total cost of the spicy noodles and the spicy rice is £17.

Method 2 Use a model.

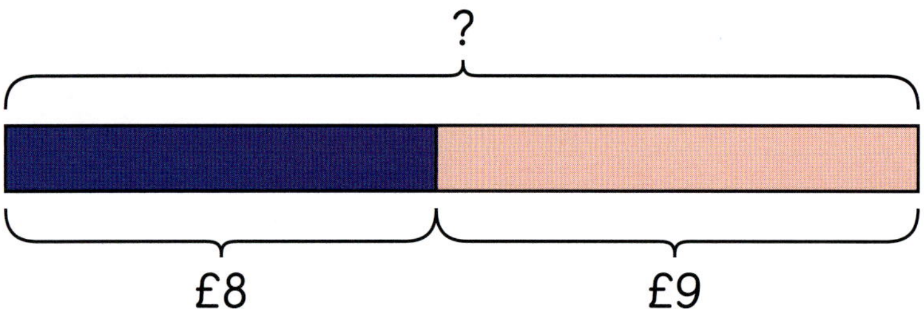

£8 + £9 = £17

The total cost of the spicy noodles and the spicy rice is £17.

2 Charles and his dad buy some sauces.

(a) Charles' dad buys sweet chilli sauce and chilli sauce.
How much does he pay altogether?

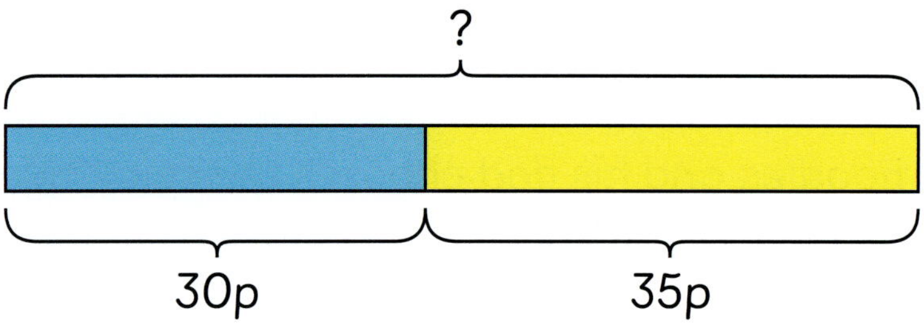

30p + 35p = 65p

Charles' dad pays 65p altogether.

(b) Charles buys soy sauce.
How much do Charles and his dad pay for sauces altogether?

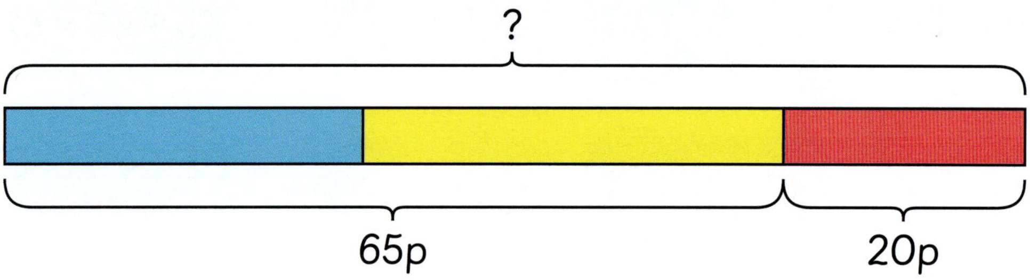

65p + 20p = ____ p

Charles and his dad pay ____ p for sauces altogether.

Guided Practice

The prices of some fruit are shown below.

Find the total cost of the following fruit.

1. 

2. 

3. 

4. 

Complete Worksheet **8**: pages **31–32**

Calculating Change

Lesson 9

Explore

Ruby had £80.
She spent £35 at a theme park.
She spent £15 at a climbing centre.
How much money did she have left?

Master

1 (a) How much did Ruby spend altogether?

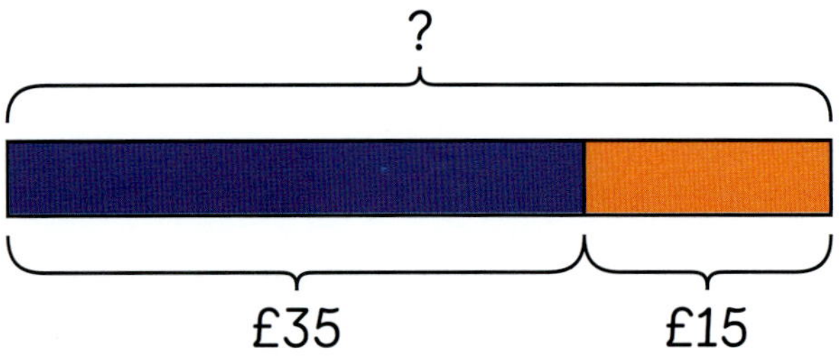

£35 + £15 = £50
Ruby spent £50 altogether.

(b) How much did she have left?

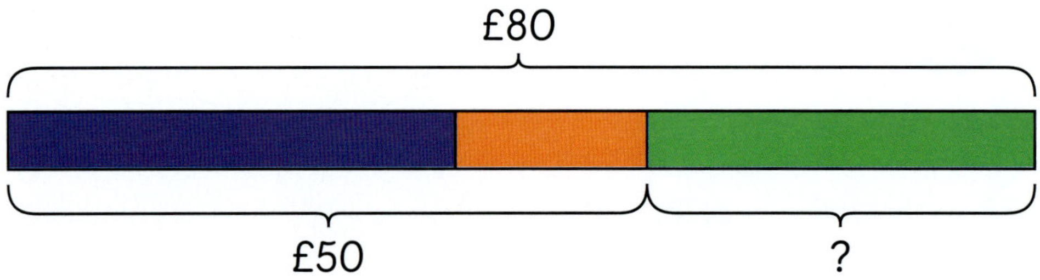

£80 − £50 = £30
Ruby had £30 left.

2 Lulu gives a twenty-pound note to a shopkeeper to pay for the following items.

How much change does Lulu get from the shopkeeper?

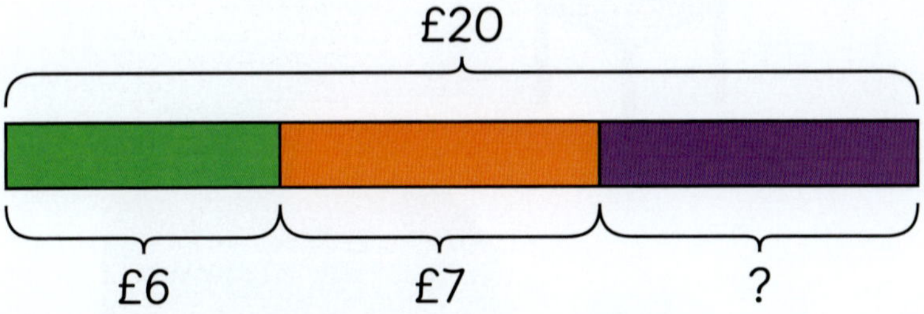

____ ○ ____ = ____

Lulu pays ____ for the  and the .

____ ○ ____ = ____

Lulu gets ____ change from the shopkeeper.

Activity Time

Partner work

You will need:

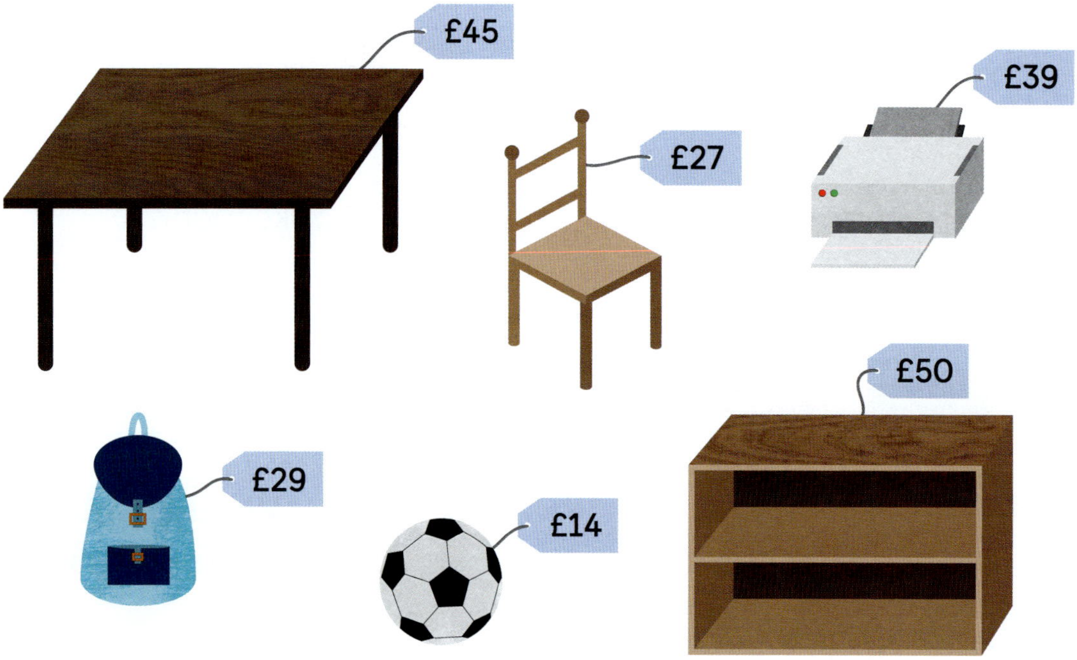

1. Choose two items to buy.
2. Work with your partner to find the total cost of the two items.
3. Pay for the two items using  or  .
4. Work with your partner to find the correct change.
5. Take turns choosing two different items to buy.

Guided Practice

Solve.

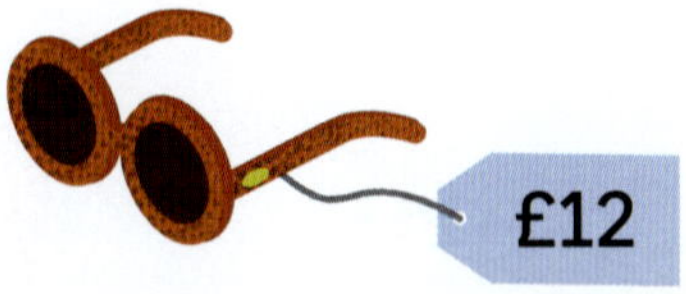

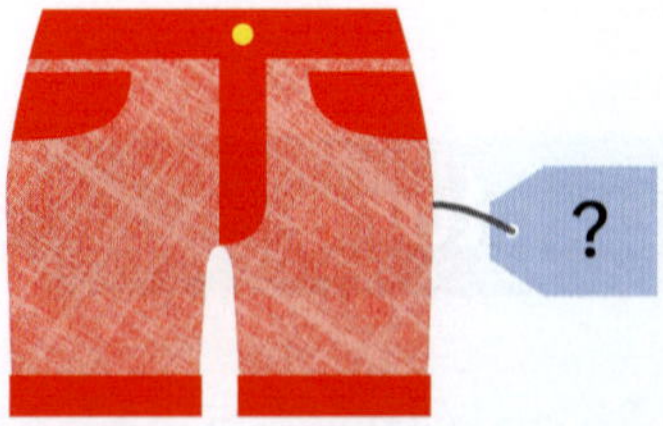

1 Hannah buys  and  using  .

How much change does she get?

2 Amira buys  ,  and  .

She pays using  and gets 

change.

How much did she pay for the 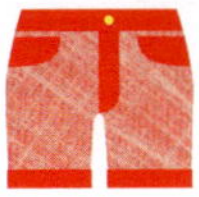 ?

Complete Worksheet **9**: pages **33–34**

Solving Word Problems

Lesson 10

Explore

Jacob buys a robot and a car.
The car costs £8 more than the robot.
How much does he spend altogether?

Master

1 Charles used a model this way.

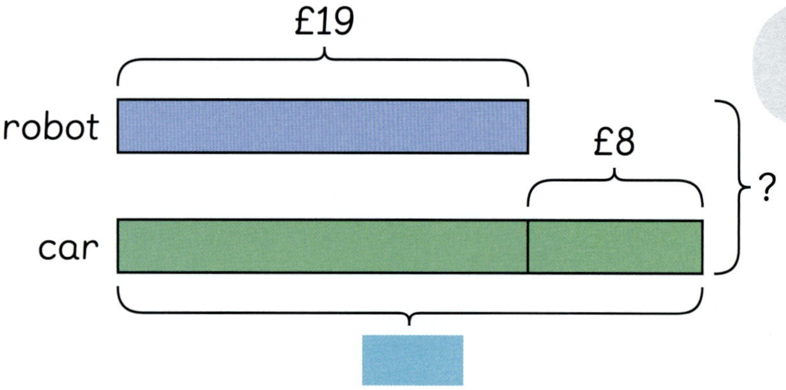

£19 + £8 = £27
The car costs £27.
£19 + £27 = £46
Jacob spends £46 altogether.

2 Ruby used a model this way.

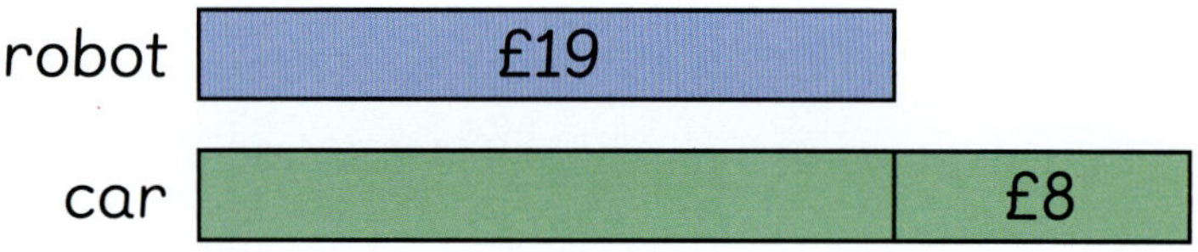

£19 + £19 + £8 = £46
Jacob spends £46 altogether.

Guided Practice

Solve.

1 The rounders bat costs £5 more than the basketball.
What is the cost of the rounders bat and the basketball altogether?

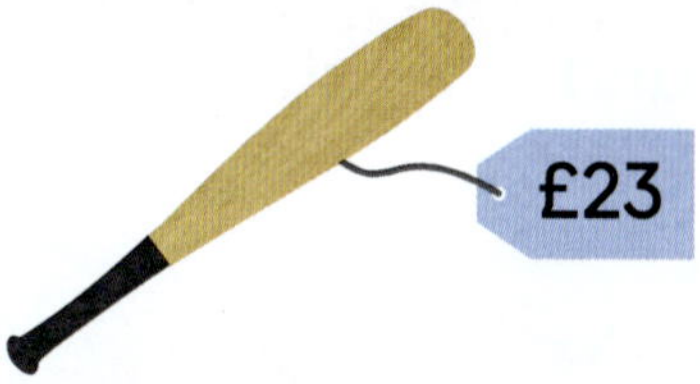

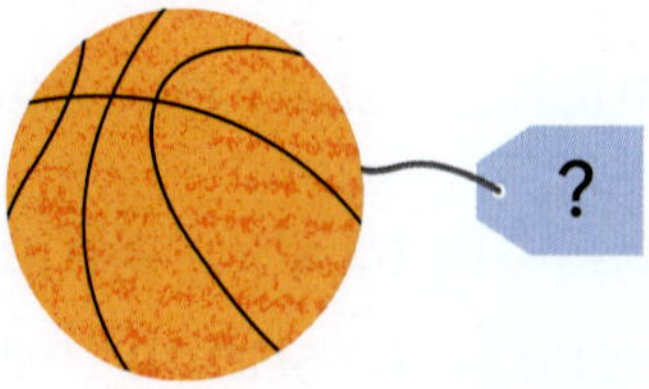

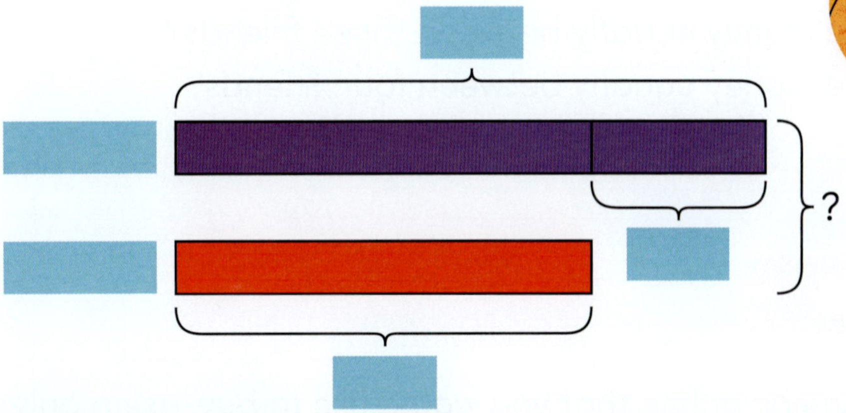

☐ ○ ☐ = ☐

The basketball costs ☐.

☐ ○ ☐ = ☐

The rounders bat and the basketball cost ☐ altogether.

2 A packet of popcorn costs 56p.
A packet of crisps costs 18p less than the packet of popcorn.
What is the total cost of the packet of popcorn and the packet of crisps?

Complete Worksheet **10**: pages **35–36**

Mind Challenge

 and  share this money equally.

How much do they get each?

Is it possible to share the money equally between three friends?
Is it possible to share the money equally between four friends?
Explain your answers.

My Maths Journal

Find an item in a magazine or online that you would like to buy, using only 2 notes and 5 coins.

Draw the item and the money in your journal.

Review

I can...

- ☐ name coins and notes.
- ☐ count to find an amount of money.
- ☐ show amounts of money in different ways.
- ☐ exchange coins and notes.
- ☐ compare different amounts of money.
- ☐ calculate change.
- ☐ solve word problems involving money.

How can we sort these shapes into two groups?

Chapter 11
2D Shapes

Identifying Sides

Lesson 1

Explore

Ravi cut out some shapes.

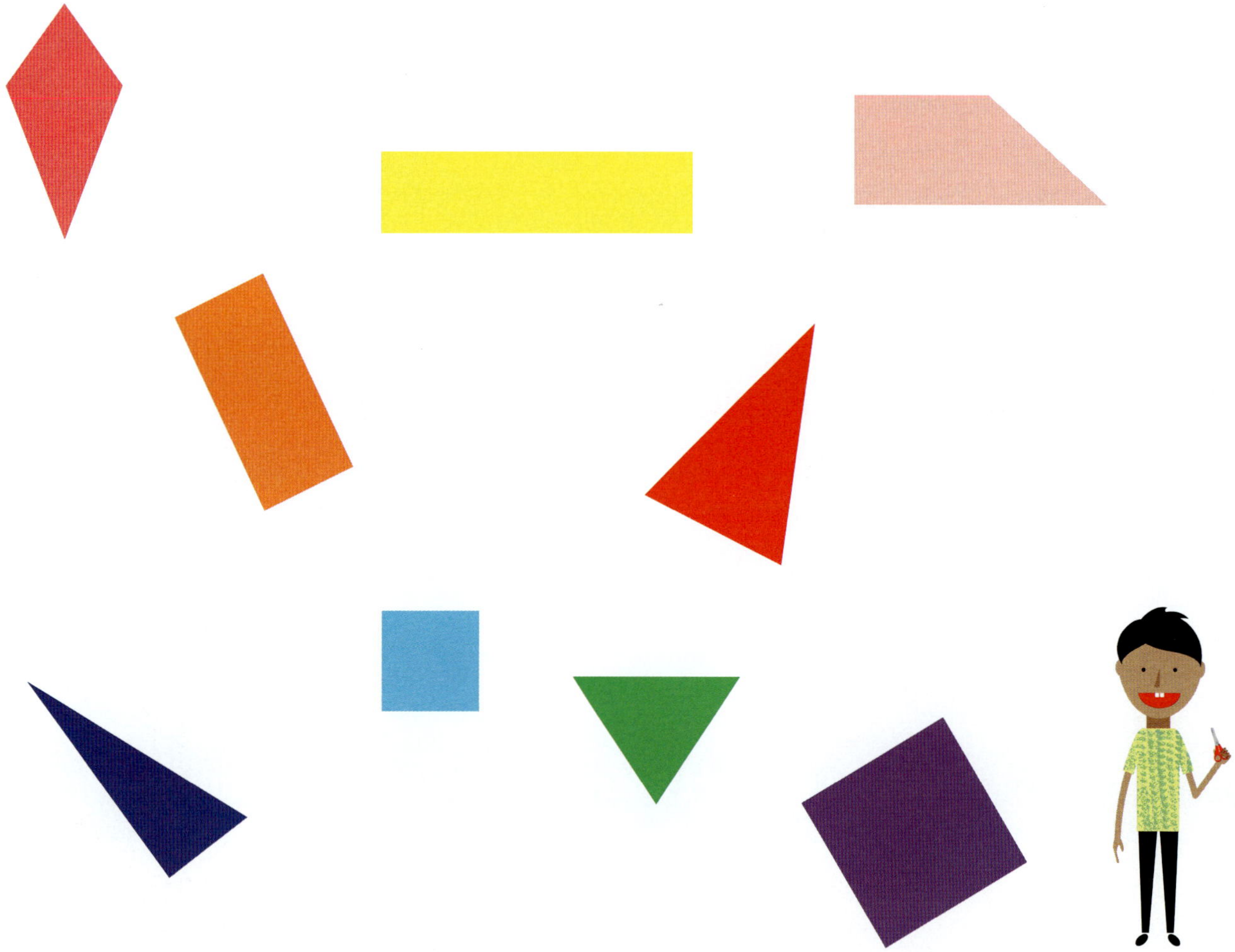

How can we describe the shapes he has cut out?

Master

1

These are triangles.
Every triangle has **3** sides.

2

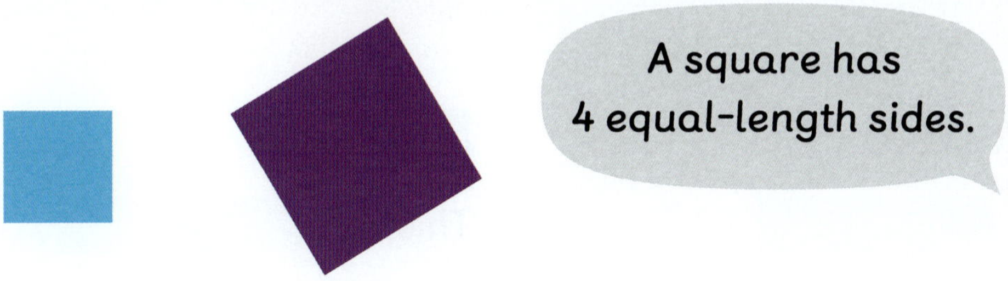

Every square has **4** equal sides.

3

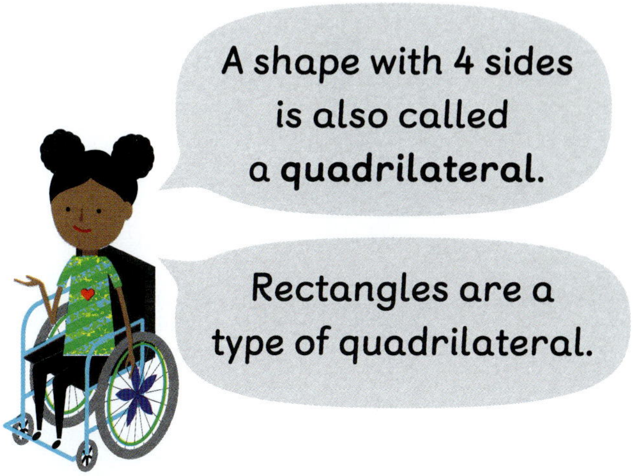

These are rectangles.
Every rectangle has ______ pairs of equal-length sides.

Guided Practice

1 Count the number of sides on each shape.
Write the name of each shape.

(a)

____ sides

This is a ________.

(b)

____ sides

This is a ________.

(c)

____ sides

This is a ________.

(d)

____ sides

This is a ________.

2 Count the number of sides on each shape.

(a)

____ sides

(b)

____ sides

(c)

____ sides

Complete Worksheet **1**: pages **43–44**

Identifying Vertices

Lesson 2

Explore

This is a group of shapes called polygons.

This is a group of shapes that are not polygons.

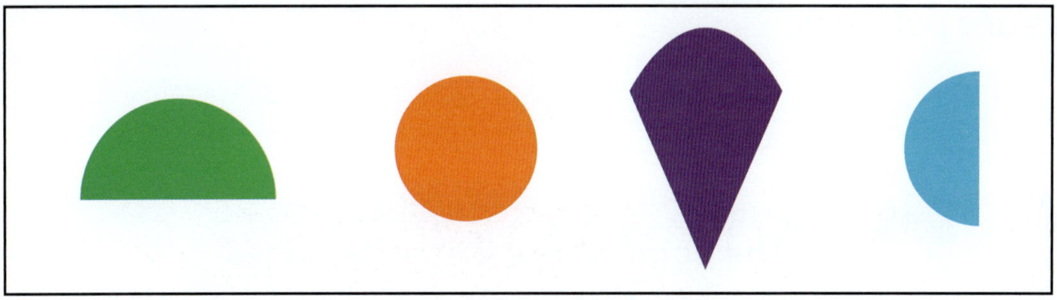

What can you say about the shapes in each group?

Master

1

Not all of the sides on these shapes are straight.
These shapes are not polygons.

2

All of the sides on each of these shapes are straight.
These shapes are polygons.

3

The plural of vertex is **vertices**.
Every triangle has 3 straight lines.
Every triangle has 3 vertices.

4 How many sides do polygons have?
How many vertices do polygons have?

This polygon has 4 sides.
It has 4 vertices.

This polygon has 3 sides.
It has ____ vertices.

This polygon has ____ sides.
It has ____ vertices.

This polygon has ____ sides.
It has ____ vertices.

Activity Time

Partner work

You will need:

1. Fold a square piece of paper like this.

2. Cut along the lines you folded.
3. Arrange the pieces to make different shapes.
4. Talk to your partner about the number of sides and vertices in the shapes you make.

Can you name the shapes?

Guided Practice

1 Name the polygon and fill in the blanks.

(a)

______

___ sides

___ vertices

(b)

______

___ sides

___ vertices

(c)

______

___ sides

___ vertices

(d)

______

___ sides

___ vertices

2 Find the number of sides and vertices for each polygon.

(a)

___ sides

___ vertices

(b)

___ sides

___ vertices

(c)

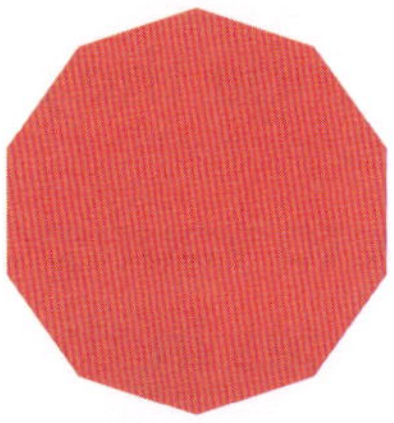

___ sides

___ vertices

What do you notice about the number of sides and the number of vertices?

Complete Worksheet **2**: pages **45–46**

Identifying Lines of Symmetry

Lesson 3

Explore

Hannah folds this square exactly in half.
Can she do the same with the other shapes?

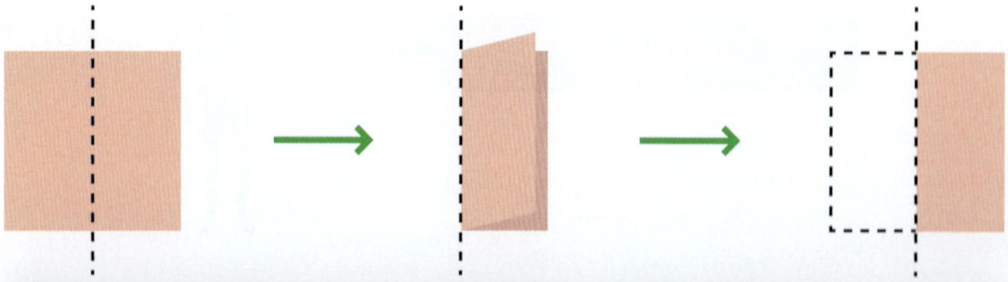

Master

1  folds the  in two ways.

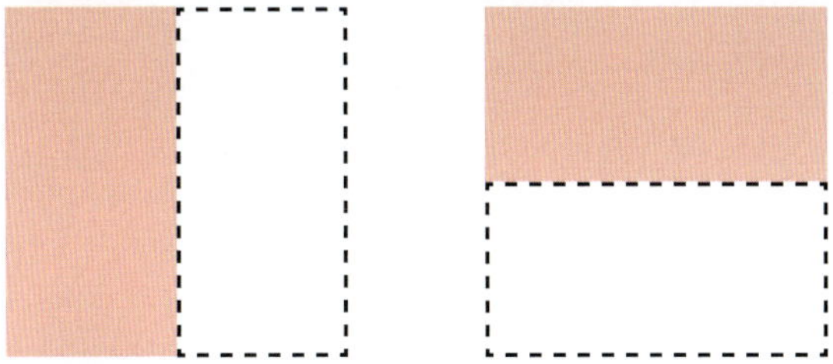

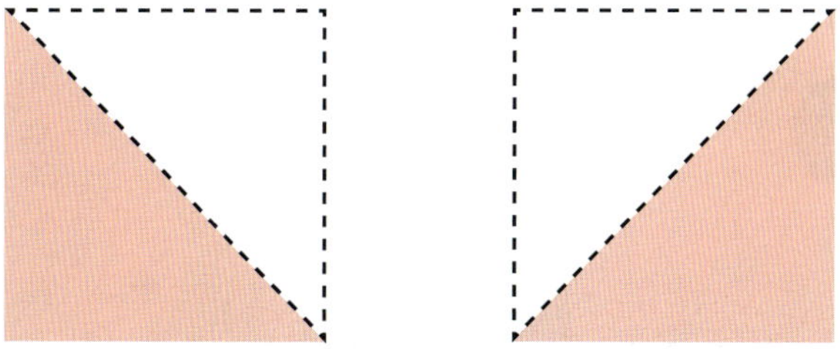

2  folds the  this way.

3  folds the  like this.

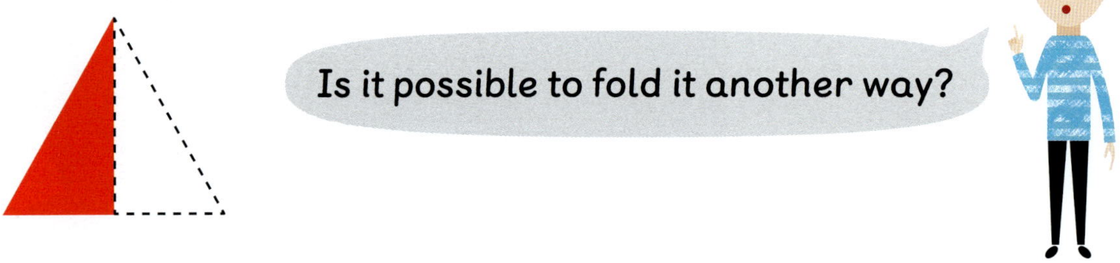

4 The line where a shape is folded so that one half overlaps the other half exactly is called the **line of symmetry**.

If a shape has a line of symmetry, we say it is **symmetrical**.

5

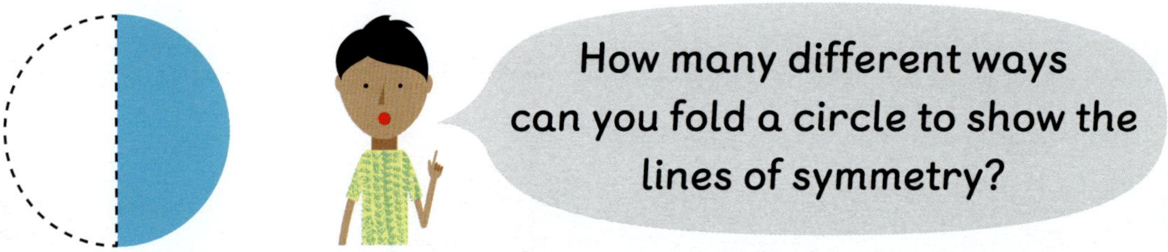

6

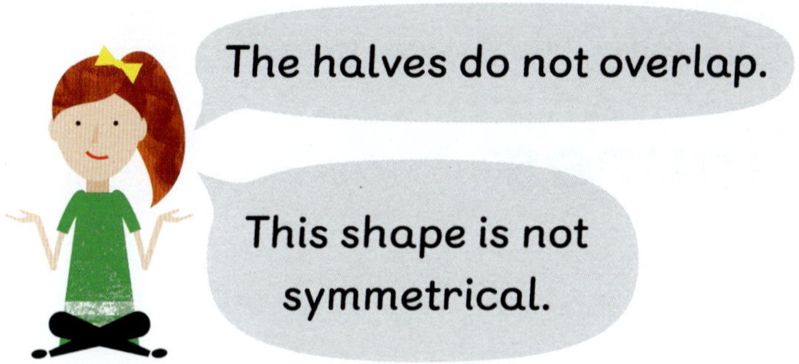

Activity Time

Group work (3 to 4 pupils)

You will need:

1. Fold each shape so one half covers the other half exactly.
2. Find all the possible lines of symmetry in the shape.
3. Point out the lines of symmetry to your classmates.
4. Your classmates can check if you are correct by folding the shape.

Guided Practice

Sort the shapes into two groups.

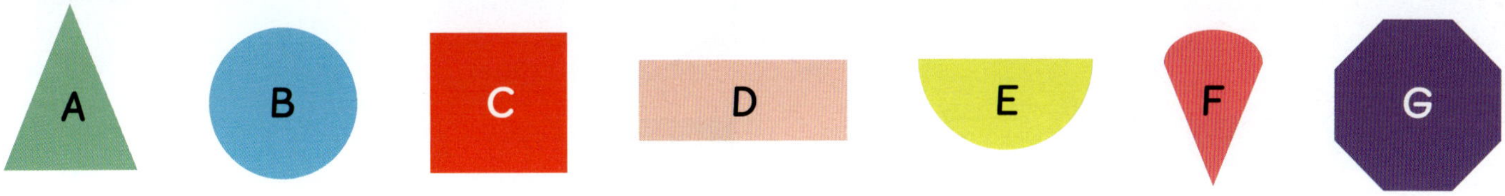

1 line of symmetry	More than 1 line of symmetry

Complete Worksheet **3**: pages **47–48**

Making Figures

Lesson 4

Explore

The children want to make symmetrical figures using some shapes.

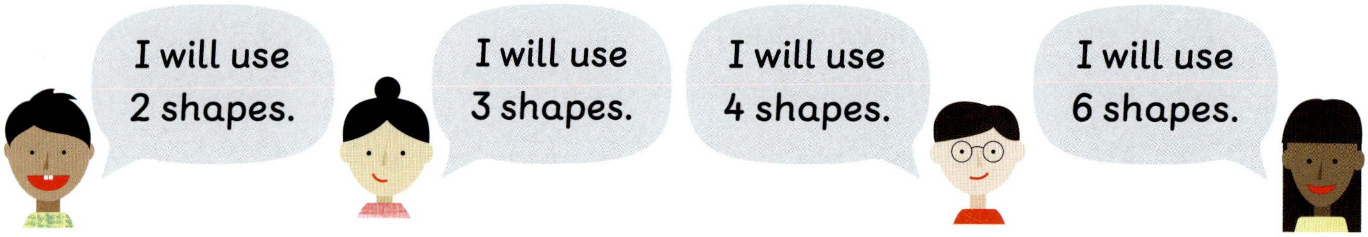

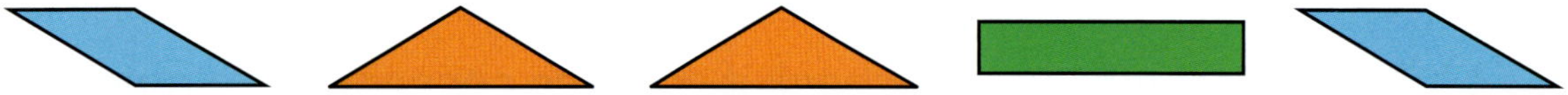

Is this possible?

Master

1 Ravi makes this figure using 2 shapes.

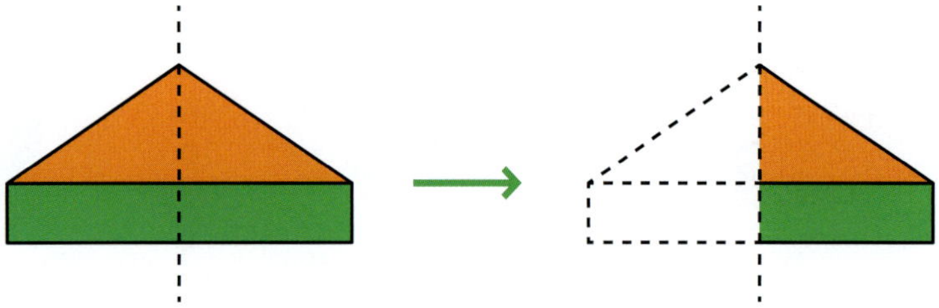

line of symmetry

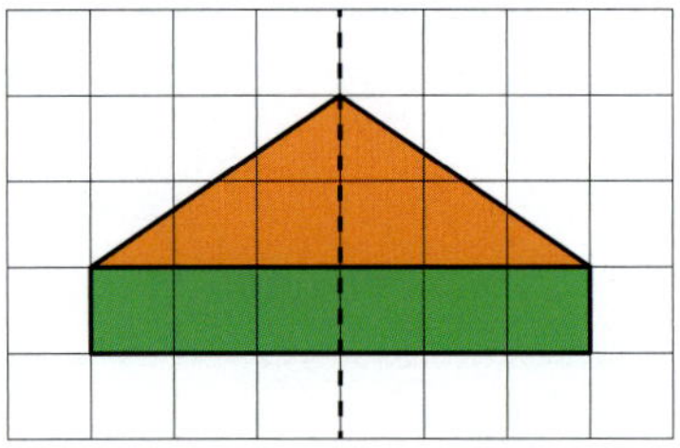

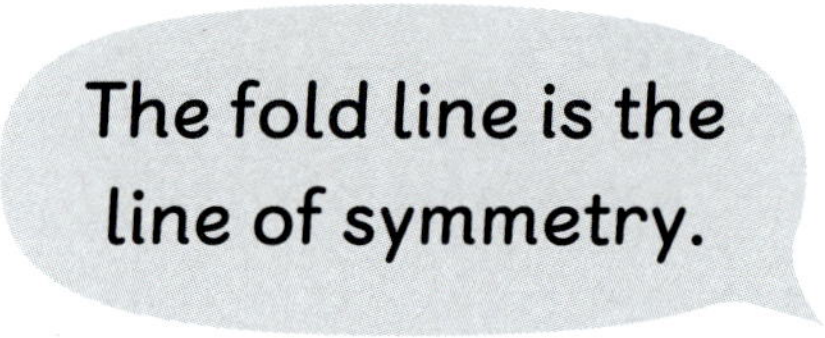

Ravi's figure has a line of symmetry.
His figure is symmetrical.

2 Ruby makes this figure using 3 shapes.

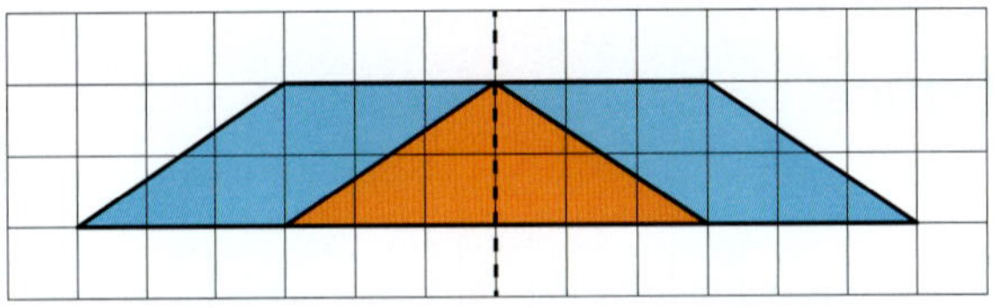

Ruby's figure has a line of symmetry.
Her figure is symmetrical.

3 Charles makes this figure using 4 shapes.

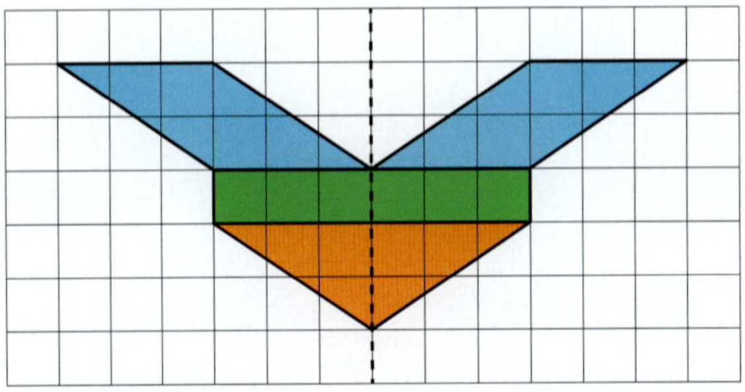

Charles' figure has a line of symmetry.
His figure is symmetrical.

4 Lulu makes this figure using 6 shapes.

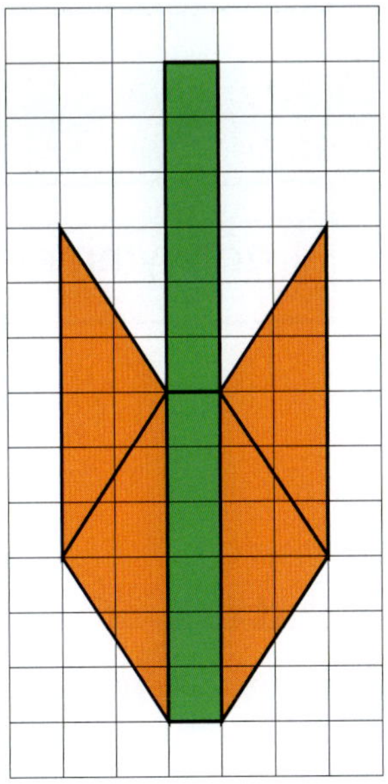

Lulu's figure has a line of symmetry.
Her figure is symmetrical.

Guided Practice

1 Draw a line of symmetry on each figure.

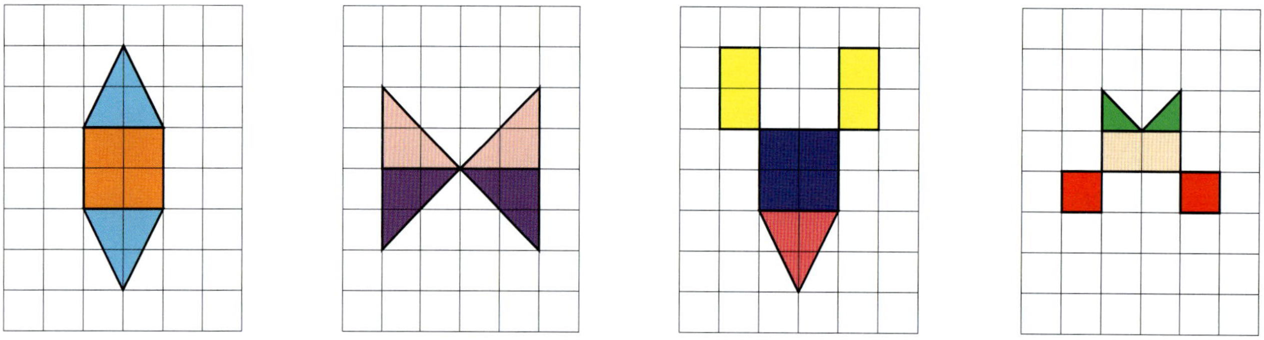

A B C D

2 Match.

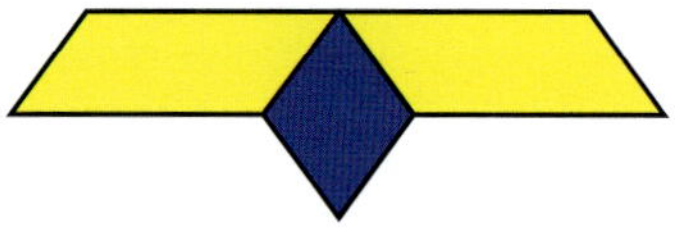

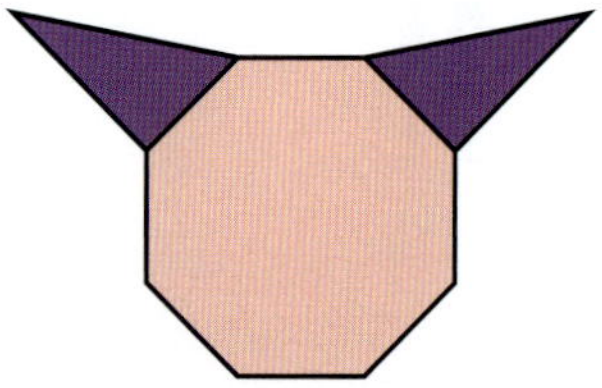

symmetrical	
not symmetrical	

Can you find the lines of symmetry on the symmetrical figures?

Complete Worksheet 4: pages **49–50**

Sorting Shapes

Lesson 5

Explore

Emma sorted these shapes into 2 groups.

How did she sort them?

Master

1

I sorted them into polygons and shapes that are not polygons.

Polygons	Not polygons

2 How did Sam sort the shapes?

I sorted them by the number of straight sides.

3 straight sides	4 straight sides	More than 4 straight sides
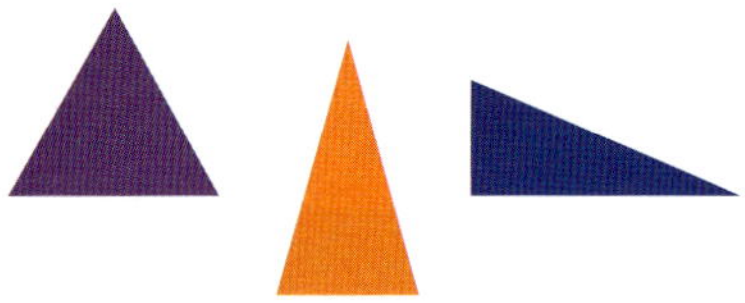		

3 How did Holly sort the shapes?

I sorted them by the number of vertices.

____ vertices	____ vertices	More than ____ vertices
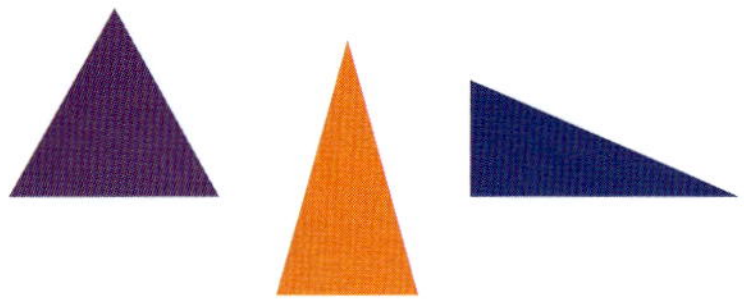		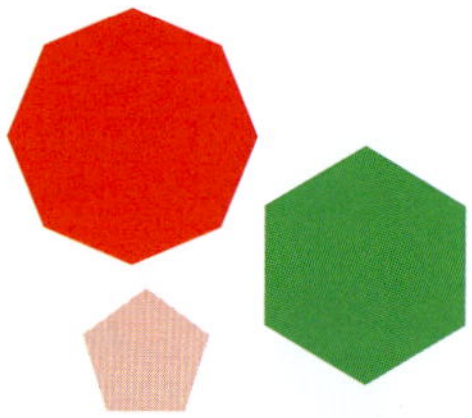

4 How did Oak sort the shapes?

Why are 2 of these shapes not in the other group?

____	____

Guided Practice

Sort the shapes into the tables.

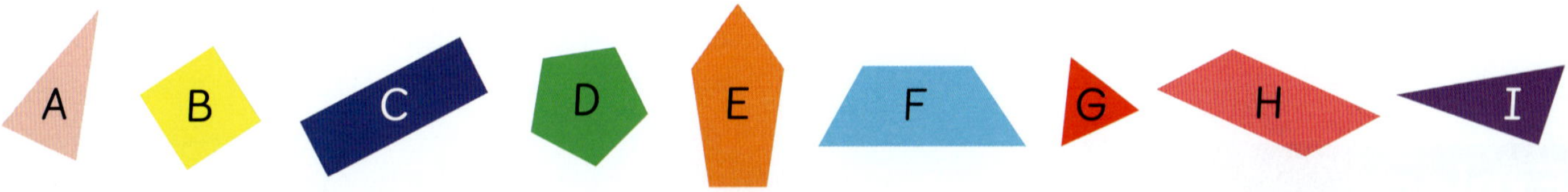

1

3 sides	4 sides	5 sides

2

3 vertices	4 vertices	5 vertices

3

Quadrilaterals	Not quadrilaterals

4

No line of symmetry	1 line of symmetry	2 lines of symmetry

Complete Worksheet **5**: pages **51–52**

Drawing Shapes

Lesson 6

Explore

Miss Fathima asks Jacob to draw a shape on square grid paper.
What shape do you think he draws?

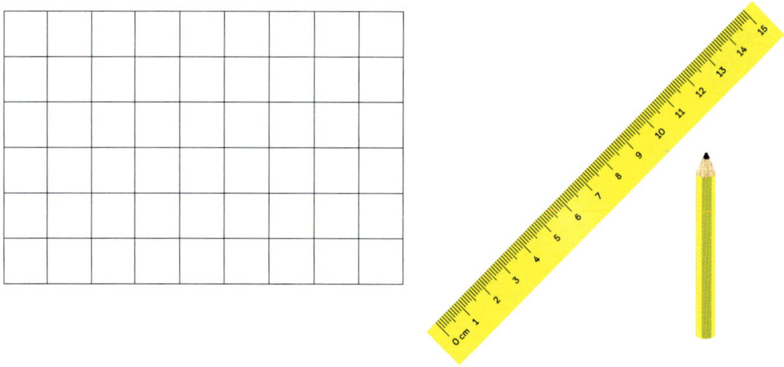

Master

1 Jacob starts drawing a shape on square grid paper.

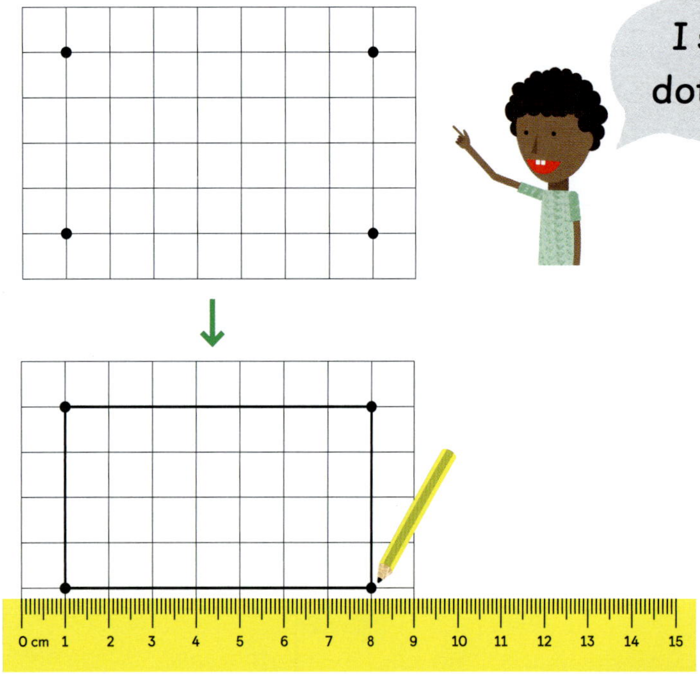

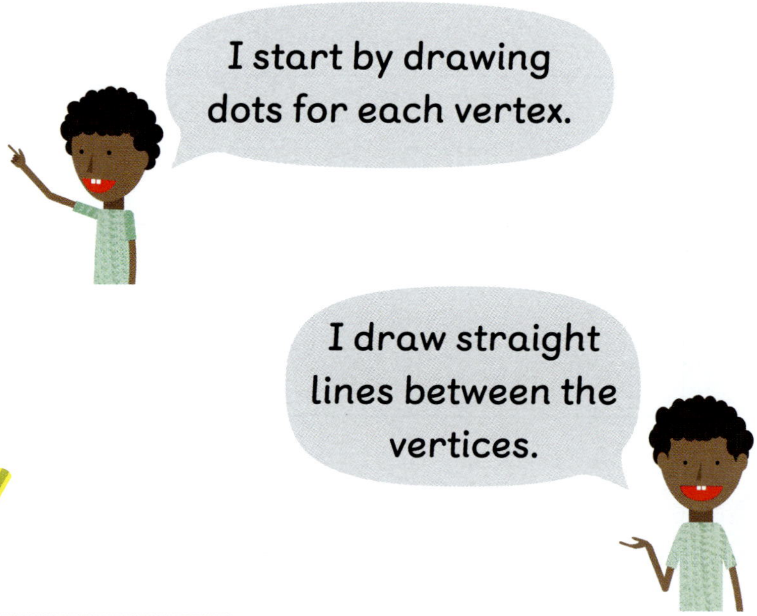

Jacob has drawn a rectangle.

2 Miss Fathima asks Hannah to draw a shape on dot paper.

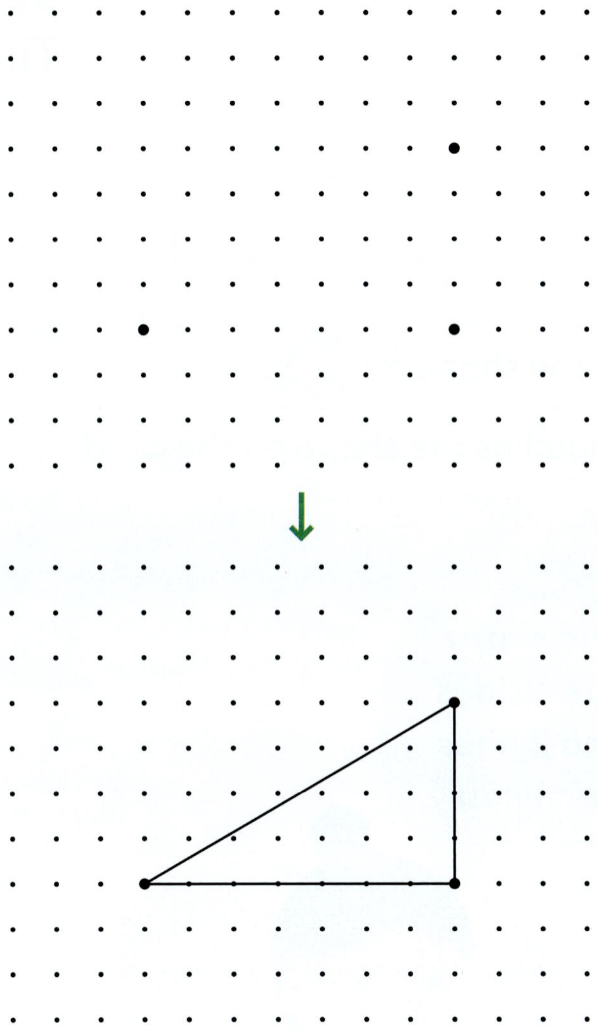

Hannah has drawn a triangle.

Activity Time

Partner work

You will need:

 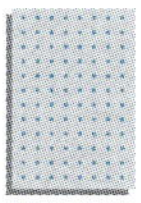

1. Pick a  and describe the shape on the card.
2. Ask your partner to draw the shape on .
3. Check the shape drawn matches the shape on the card.
4. Take turns.

Guided Practice

1 Copy each shape below onto square grid paper.

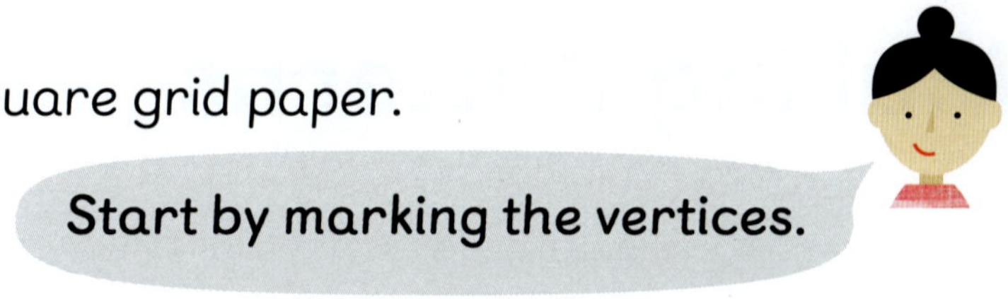

(a)

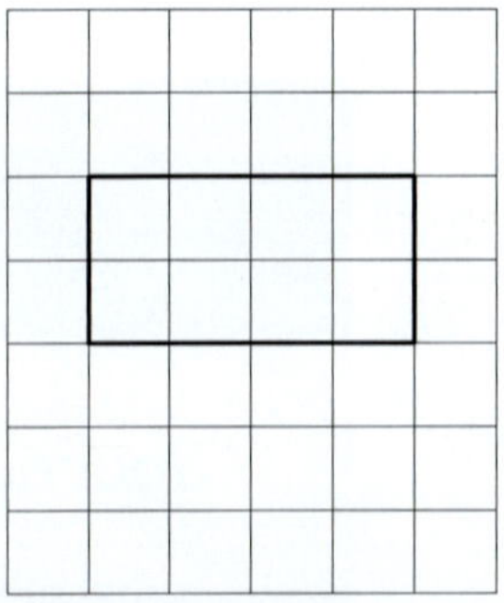

(b)

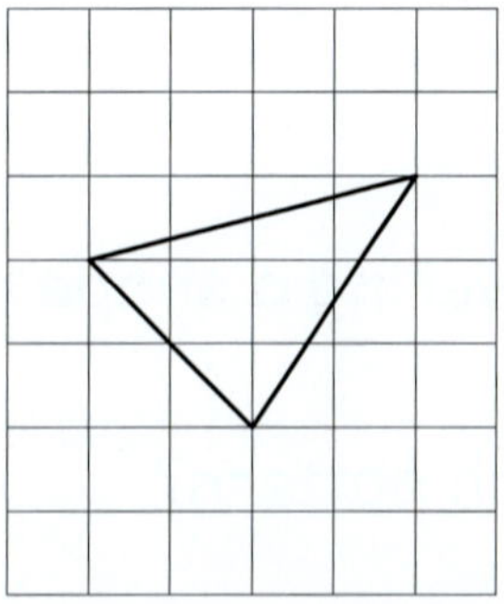

(c)

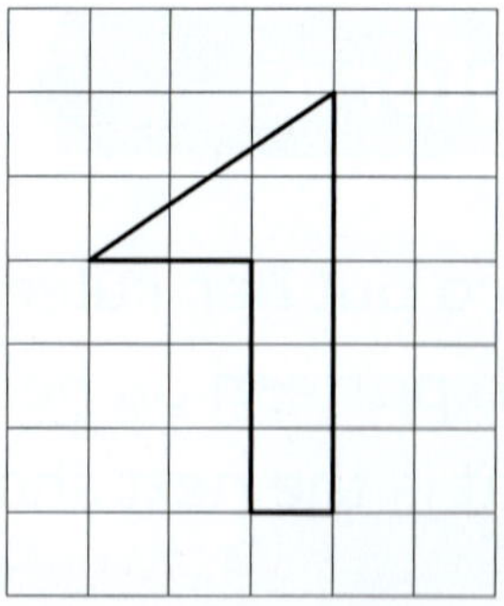

2 Copy each shape below onto dot paper.

(a)

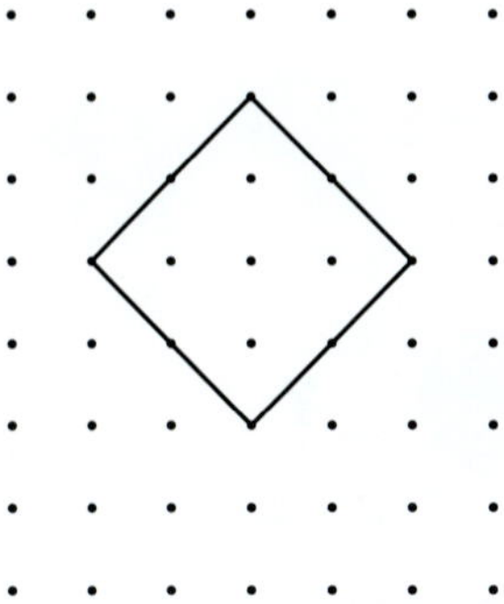

(b)

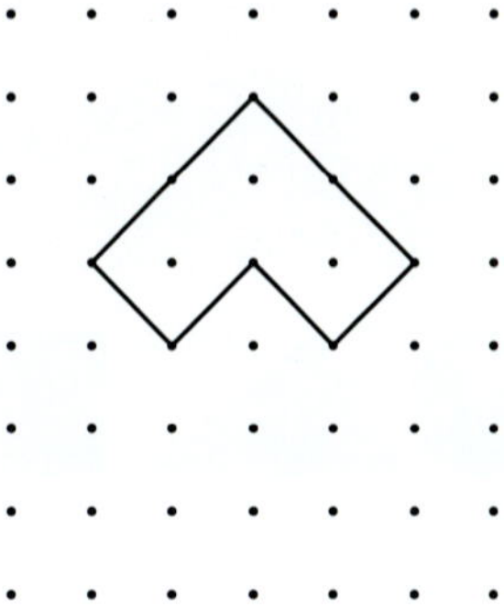

(c)

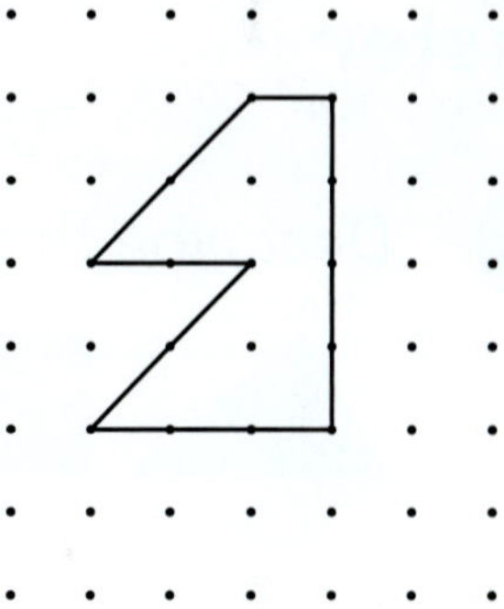

3 Draw a symmetrical shape with 5 sides.

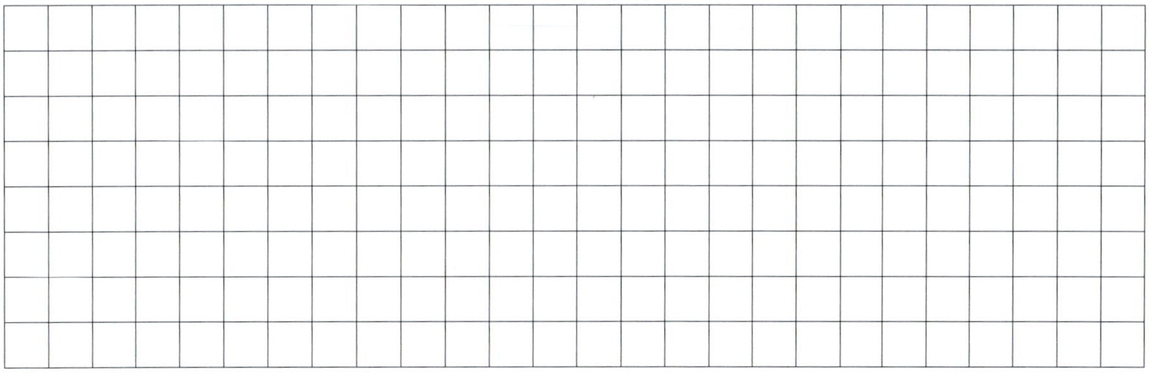

Complete Worksheet **6**: pages **53–54**

Making Patterns

Lesson 7

Explore

Amira put her ruler down covering a shape in each pattern on her journal.
What is the next shape in each pattern?

Master

1 Describe the pattern.

 ?

Triangle, rectangle, circle

3 shapes are repeated in the pattern.
The **shape** changes in the pattern.
The next shape in the pattern is  .

2 What is the next shape in this pattern?

 ?

The shape is repeated in the pattern.
The **orientation** of the shape changes.
The next shape in the pattern is  .

How does the shape change?

Activity Time

Partner work

You will need:

① Make a pattern using .

For example:

② Ask your partner the following questions:

(a) What changes in the pattern?

(b) What comes next in the pattern?

③ Check the answers together.

④ Repeat the activity, taking turns.

Guided Practice

Find the shapes missing in the patterns.

1 

2 

3 

Complete Worksheet **7**: pages **55–56**

Describing Patterns

Lesson 8

Explore

Miss A'liya rolls out some wallpaper.

1st 2nd 3rd 4th 5th 6th 7th 8th 9th

She can see the first 9 shapes.
What is the 10th shape in the pattern?

Master

1 orange circle purple rectangle

1st 2nd 3rd 4th 5th 6th 7th 8th 9th 10th

The 10th shape is  .

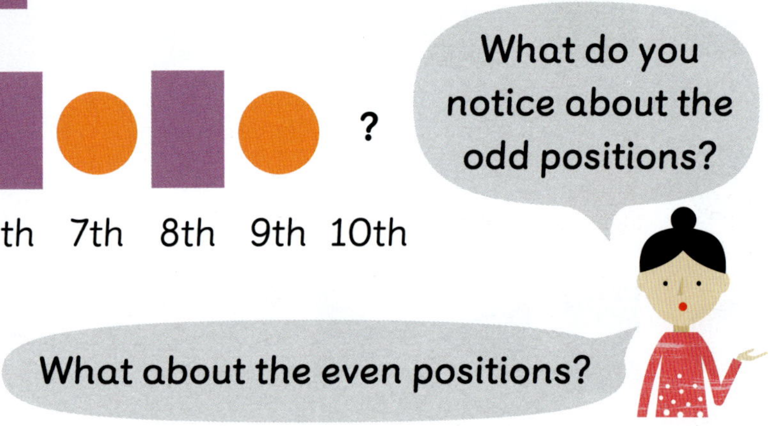

2

1st 2nd 3rd 4th 5th 6th 7th 8th 9th 10th

The 11th shape is .

The 12th shape is .

The 13th shape is .

The 14th shape is .

The 15th shape is .

The 16th shape is .

Can you find the 19th shape?

What about the 20th shape?

Guided Practice

1

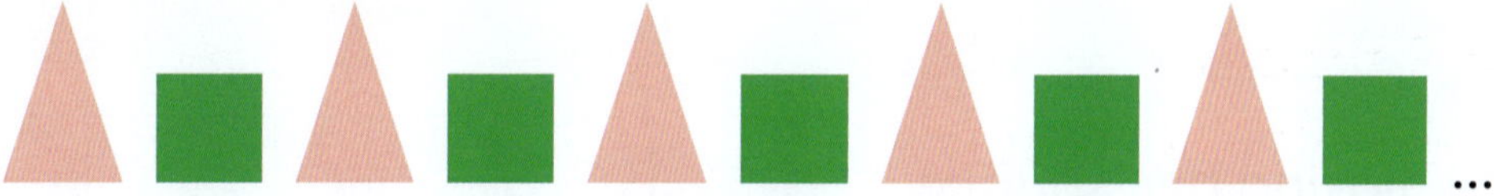

1st 2nd 3rd 4th 5th 6th 7th 8th 9th 10th

(a) The 11th shape is ______.

(b) The 12th shape is ______.

(c) The 30th shape is ______.

(d) The 35th shape is ______.

2 What is the 2nd shape?

15th 16th 17th 18th 19th 20th

3 What is the 9th shape?

2nd 3rd 4th 5th 6th

Complete Worksheet **8**: pages **57–58**

Moving Shapes

Lesson 9

Explore

Oak and Holly are playing a game.
Oak wins if she covers Holly's square with her square.
How can Oak move her square to cover Holly's square?

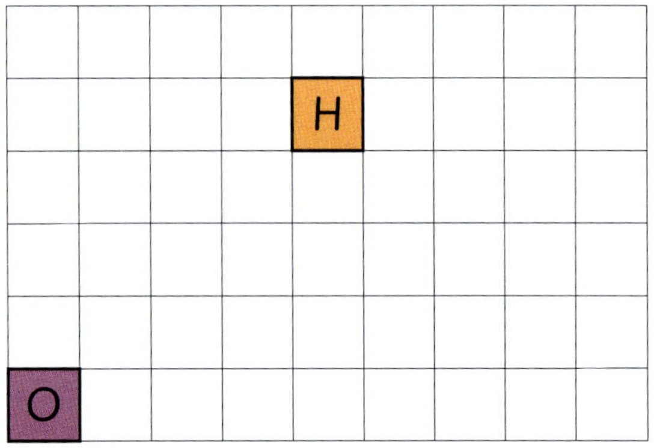

Master

1 Oak starts by moving her square 4 steps to the right.

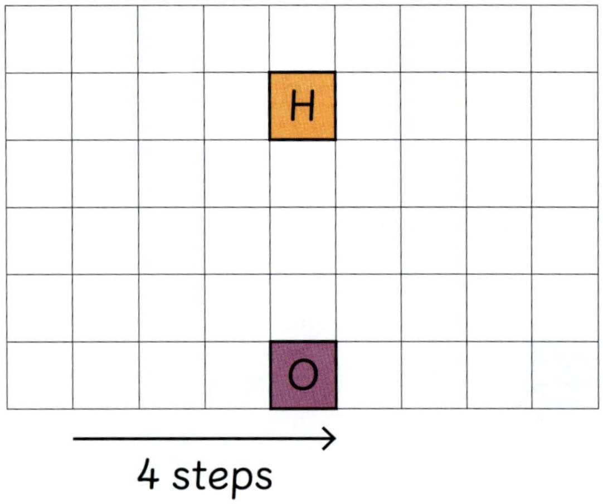

2 Oak then moves her square 4 steps up.

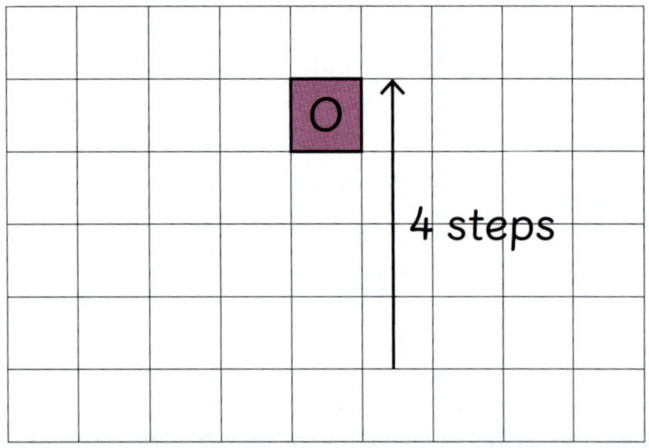

Guided Practice

Write instructions to move:

1  to 

Move ____ steps

____ (left / right).

Move ____ steps

____ (up / down).

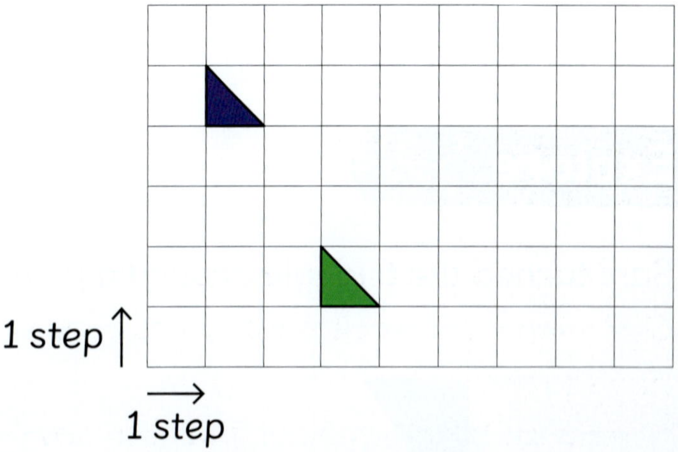

2  to 

Move ____ steps

____ (left / right).

Move ____ steps

____ (up / down).

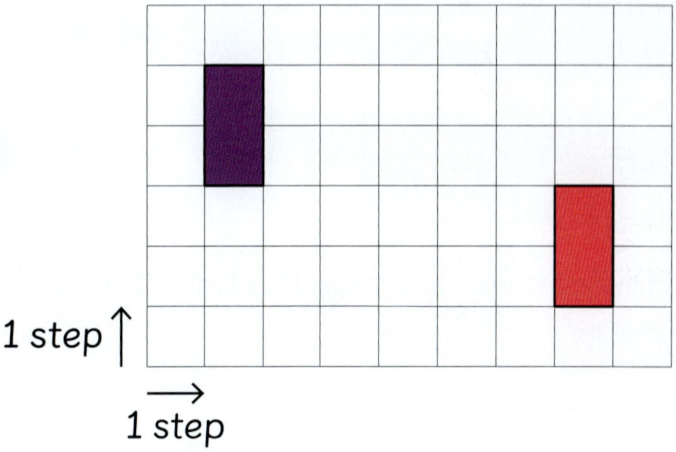

3  to 

Move ____ steps

____ (left / right).

Move ____ steps

____ (up / down).

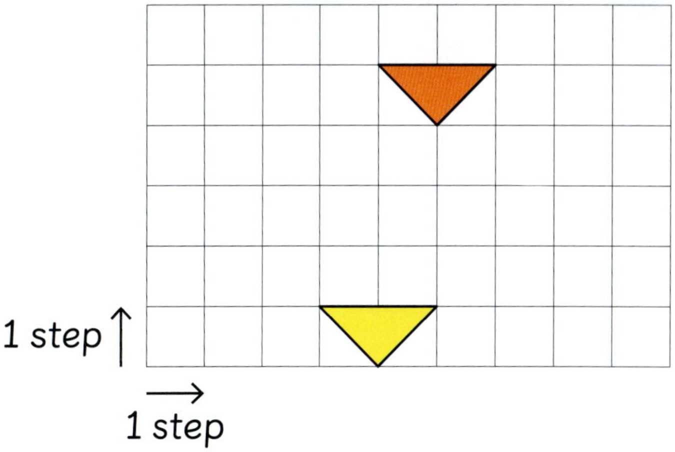

Complete Worksheet **9**: pages **59–60**

Turning Shapes

Lesson 10

Explore

Sam turned the triangle around a point.

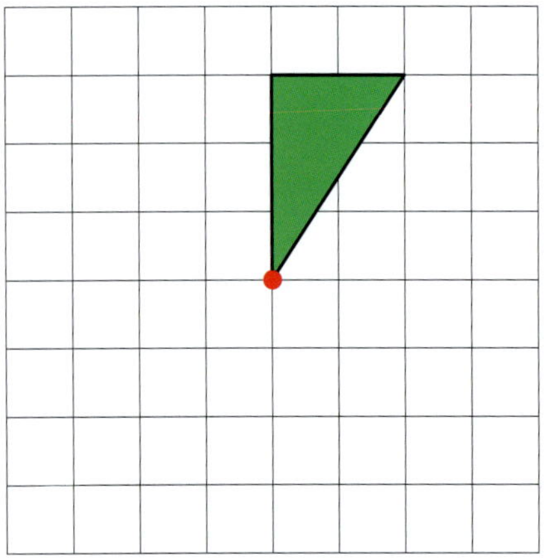

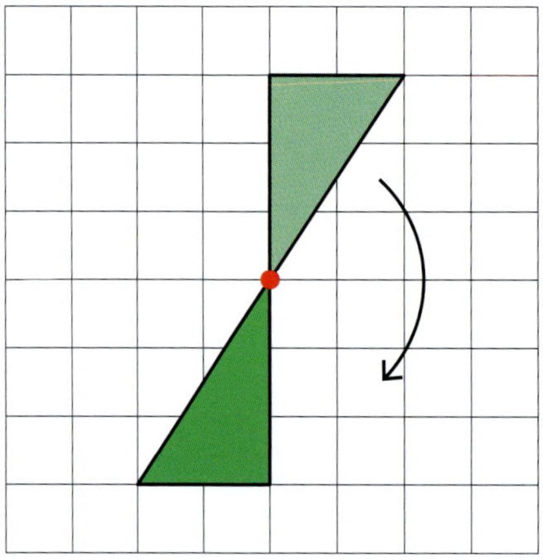

How can we describe the turn it made?

Master

1

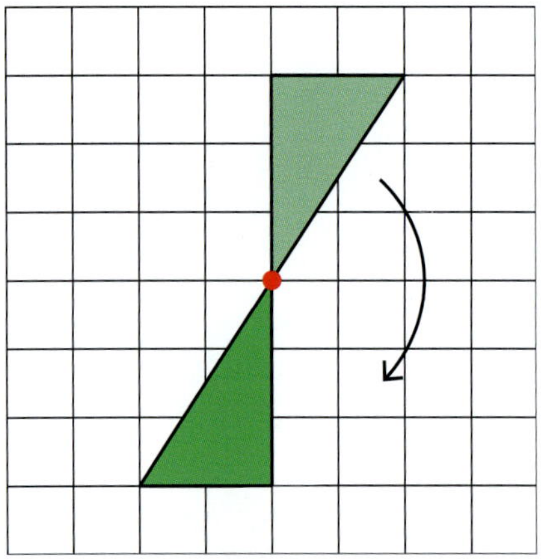

The shape is turned clockwise.
It made half a turn.
The shape made half a turn clockwise.

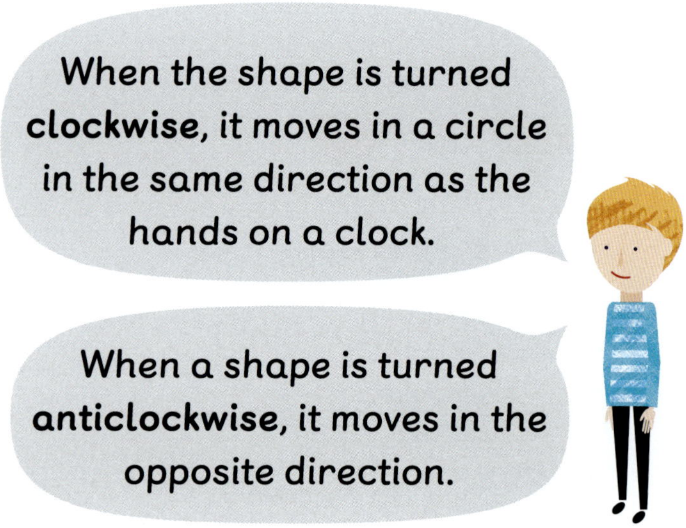

2

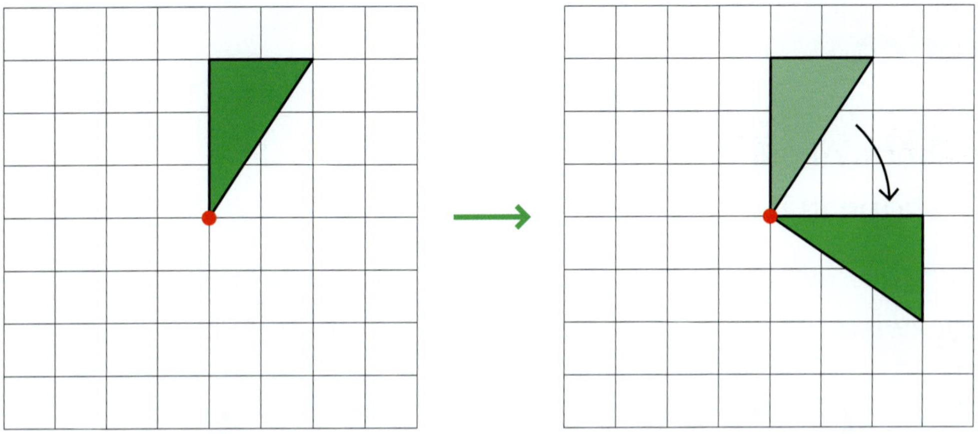

Turn the shape clockwise by a quarter turn.

3

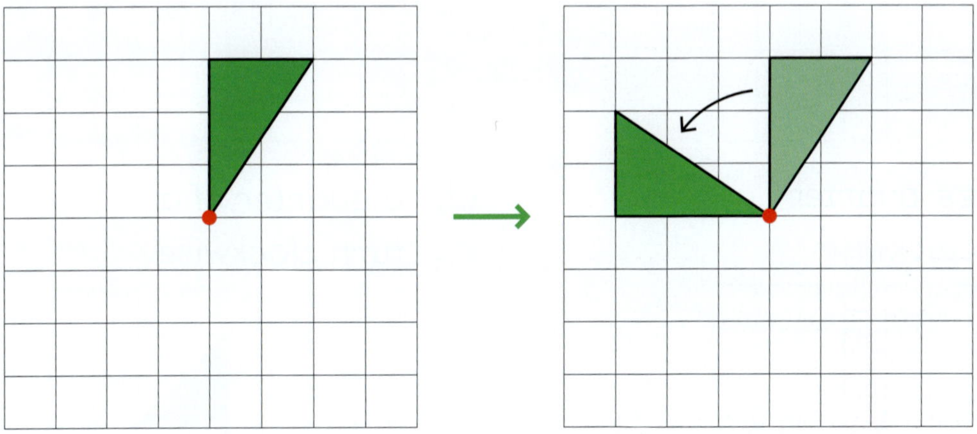

Turn the shape anticlockwise by a quarter turn.

4

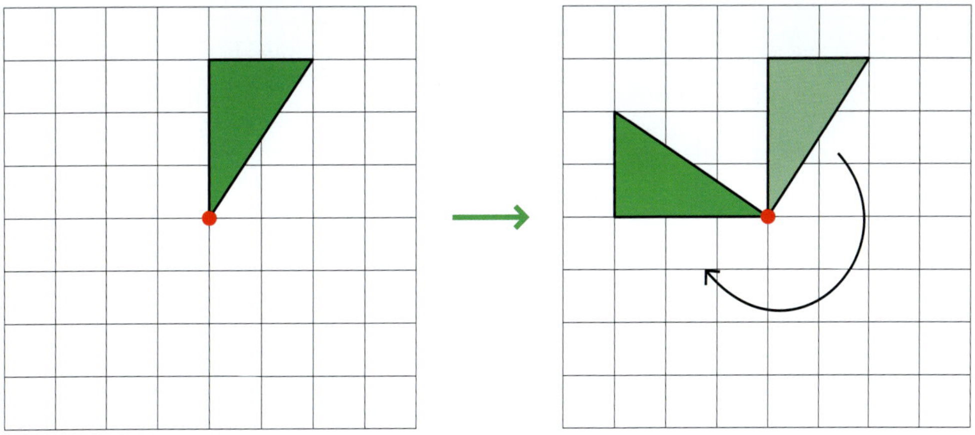

Turn the shape clockwise by a three-quarter turn.

Guided Practice

1 Turn each shape around the point as described below. Draw each shape in its new position.

(a) a three-quarter turn anticlockwise

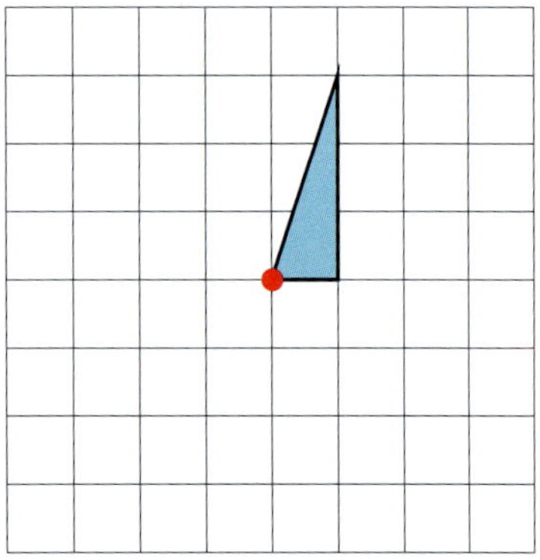

(b) half a turn anticlockwise

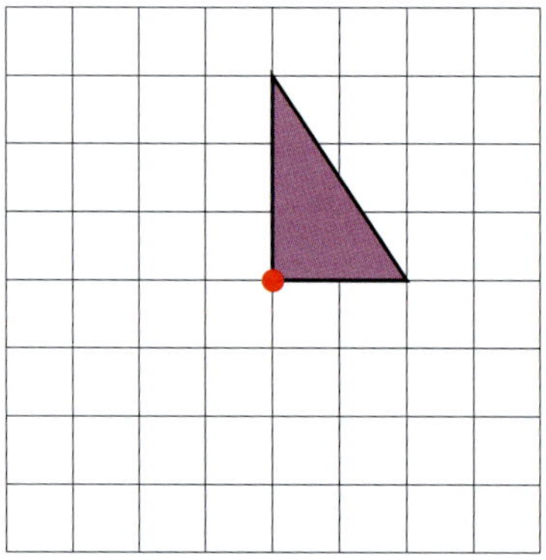

(c) a three-quarter turn clockwise

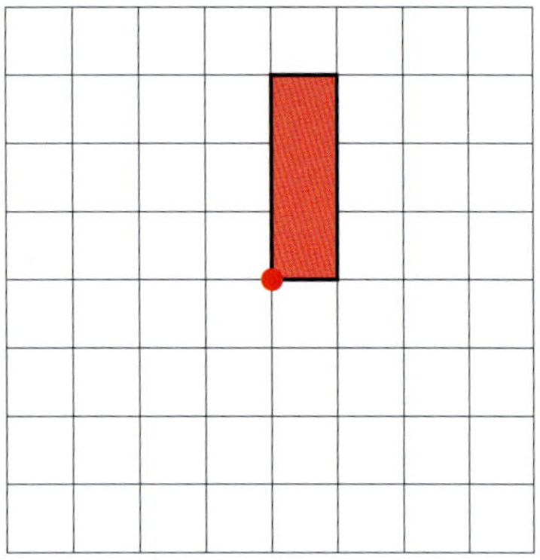

(d) a quarter of a turn clockwise

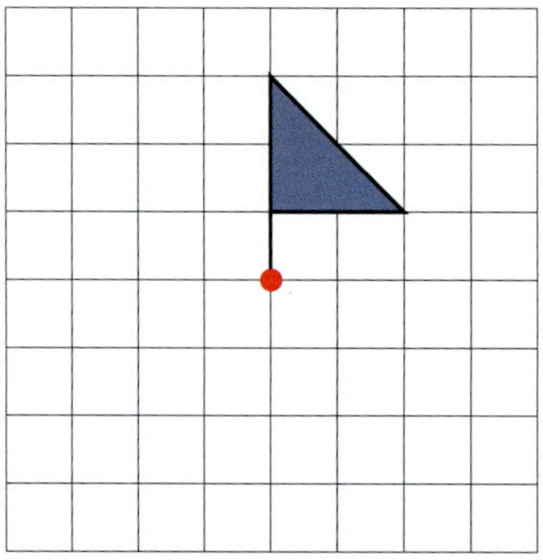

2 Describe how the green shape turns so it covers the pink shape.

(a)

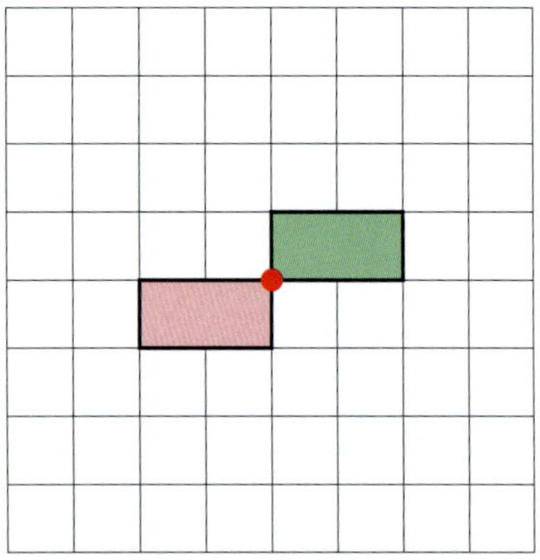

(b)

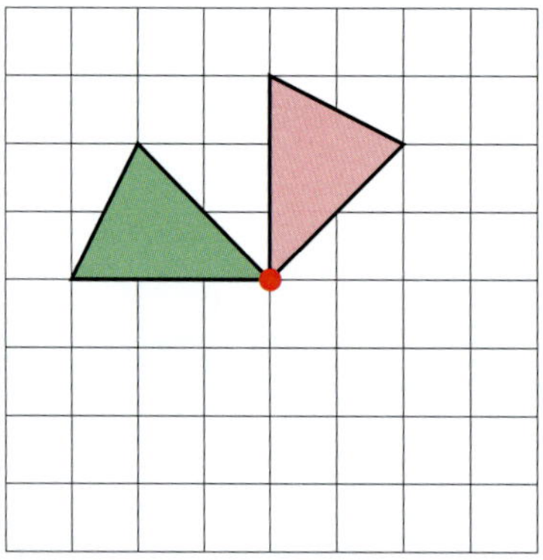

Complete Worksheet **10**: pages **61–62**

Mind Challenge

The teacher is at the front of the classroom facing the children.

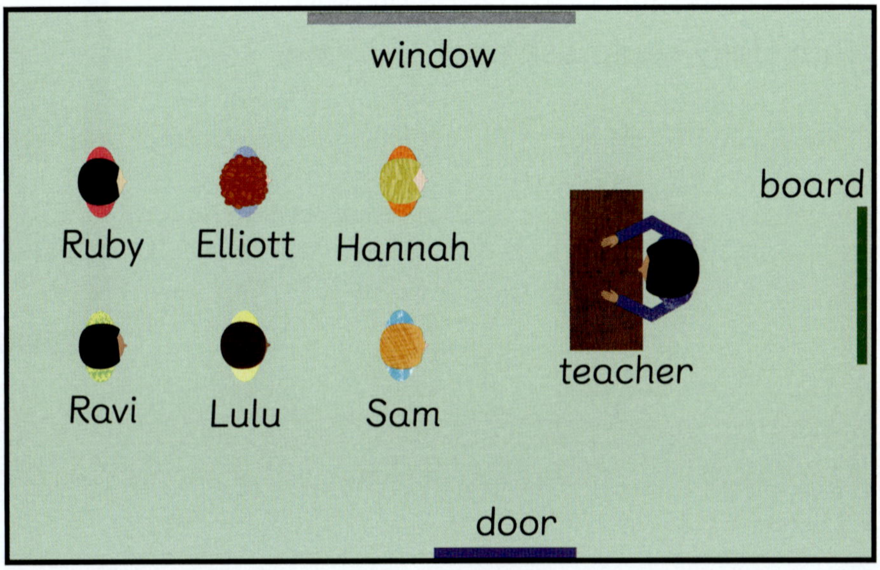

Ruby wants to face the wall with the window.
Elliott wants to face Lulu.
Hannah wants to face the back of the classroom.
Sam wants to face the wall with the door.
Ravi wants to face the wall with the board.
Lulu wants to face Elliott.

Describe the turn each child needs to make.

My Maths Journal

Holly wants to decorate the border of her picture frame. She wants to make a repeating pattern using shapes. Draw a pattern that Holly could use and describe the pattern.

I can...

Review

- ☐ name triangles, quadrilaterals and polygons.
- ☐ identify the number of sides and vertices of a shape.
- ☐ find lines of symmetry in shapes and figures.
- ☐ make figures with shapes.
- ☐ name the shapes that make up a figure.
- ☐ sort shapes.
- ☐ draw figures on a square grid and dot grid.
- ☐ make and complete patterns.
- ☐ describe patterns.
- ☐ move and turn shapes.

How can we describe the shape of each object?

Chapter 12
3D Shapes

Recognising 3D Shapes

Lesson 1

Explore

What are the shapes of the items Elliott is buying?

Master

1 

This shape is a **cuboid**.

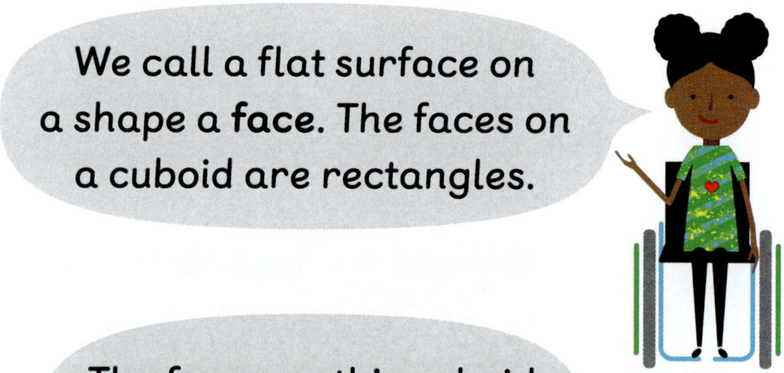

2 

This shape is also a cuboid.

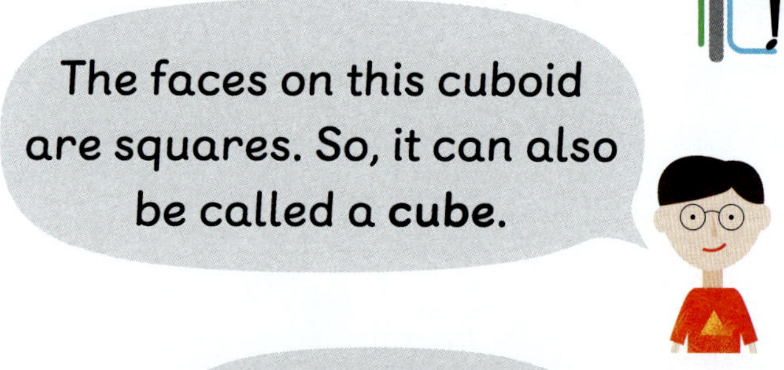

3 

This shape is a **sphere**.

4

This is a **cylinder**.
A cylinder has two circular faces and a curved surface.

Where are the faces on the cylinder?

5

This shape is a **cone**.
A cone has a circular face and a curved surface.

6

This shape is a **prism**.
The faces on this prism are triangles and rectangles.

Which shapes are best for rolling?

Which shapes are best for stacking and fit neatly together?

Guided Practice

Use the labels to name the shapes.

cube | cuboid | sphere | prism

cone | cylinder

1 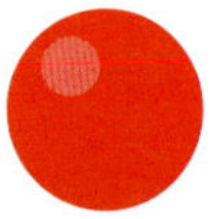

2 

3 

4 

5 

6 

7 

8 

Complete Worksheet **1**: pages **67–68**

Describing 3D Shapes

Lesson 2

Explore

How can we describe the shape of the bricks the builder is using?

Master

1

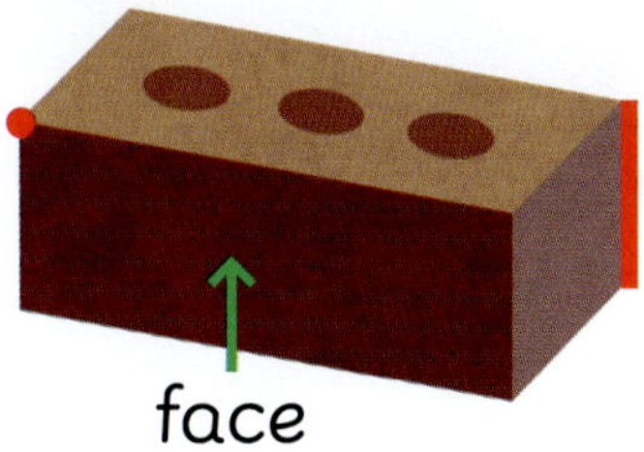

This is a cuboid.
A cuboid has 6 faces and 12 edges.
It has 8 vertices.

2

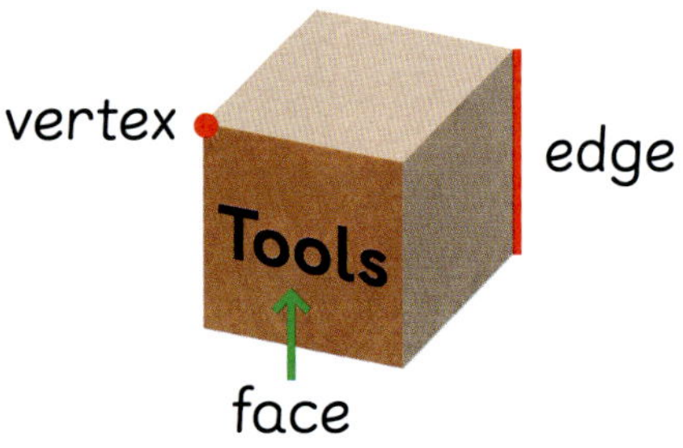

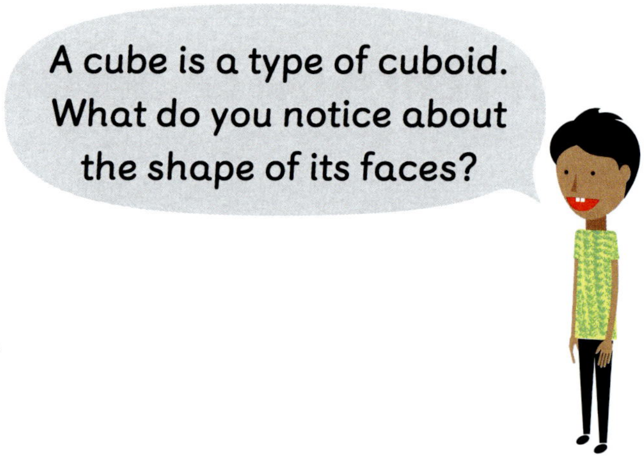

This is a cube.
A cube has 6 faces and 12 edges.
It has 8 vertices.

3

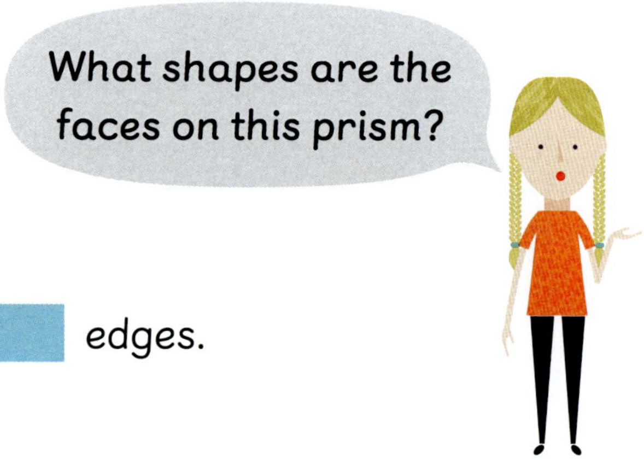

This is a prism.

This prism has ____ faces and ____ edges.

It has ____ vertices.

4

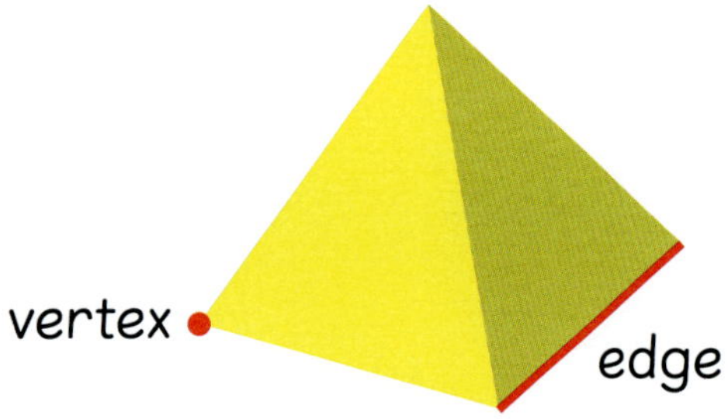

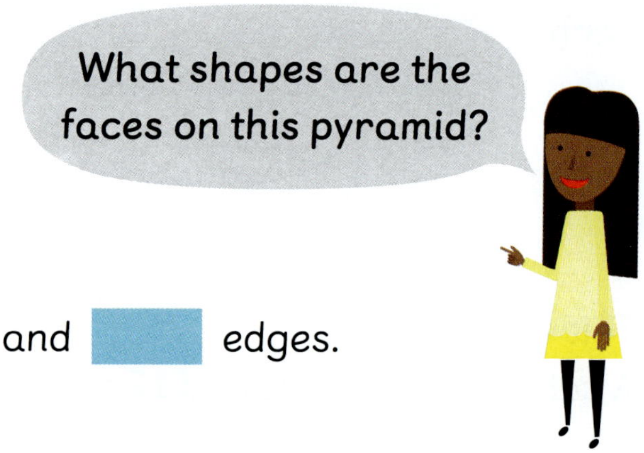

This is a pyramid.

This pyramid has ____ faces and ____ edges.

It has ____ vertices.

Guided Practice

1 Write the name of each shape.
Count the number of faces.

(a) 

______

____ faces

(b) 

______

____ faces

(c) 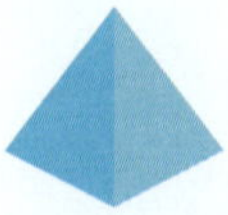

______

____ faces

(d) 

______

____ faces

2 Complete the table.

Shape	Number of edges	Number of vertices
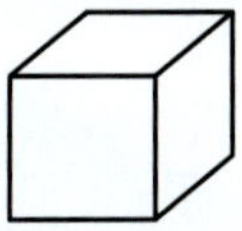		
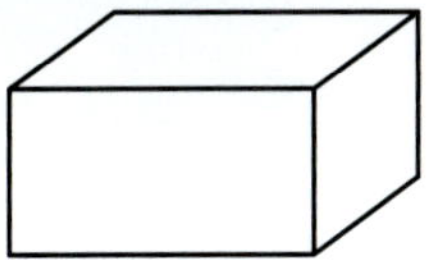		
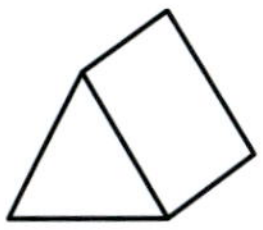		
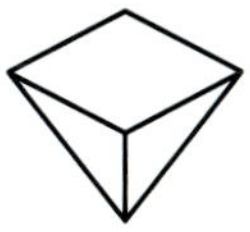		

Complete Worksheet **2**: pages **69–70**

Describing 3D Shapes

Lesson 3

Explore

Mr Nightingale wants to unfold the box so that it is flat.

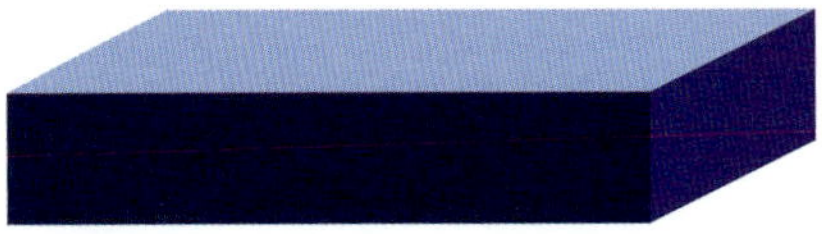

What 2D shapes will he see on the flattened box?

Master

1 Mr Nightingale unfolds the box like this.

The cuboid has 6 faces.
Each face on the cuboid is a rectangle.

2 Unfold the cube.

A cube has 6 faces.

____ faces are squares.

3 Unfold the prism.

This prism has 5 faces.

____ faces are rectangles.

____ faces are triangles.

4 Unfold the pyramid.

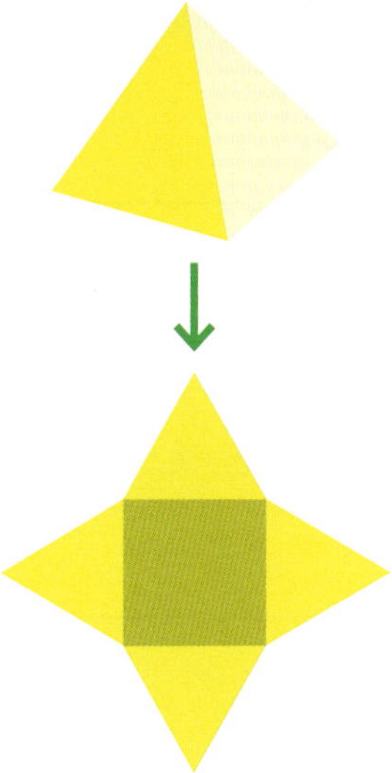

This pyramid has ____ faces.

____ face is square.

____ faces are triangles.

Guided Practice

Complete the table below.

Object	Shape(s) of faces	Number of faces
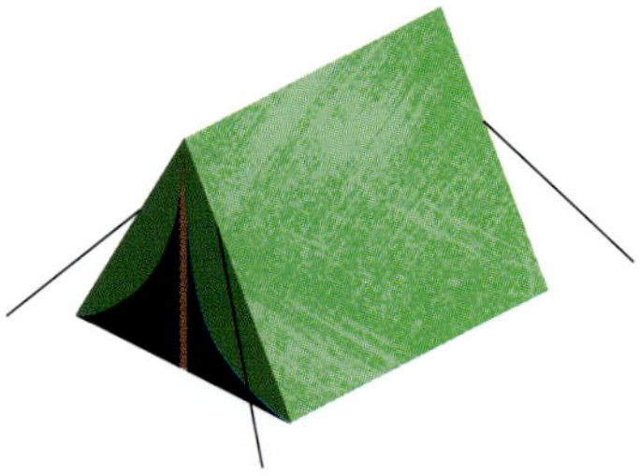		

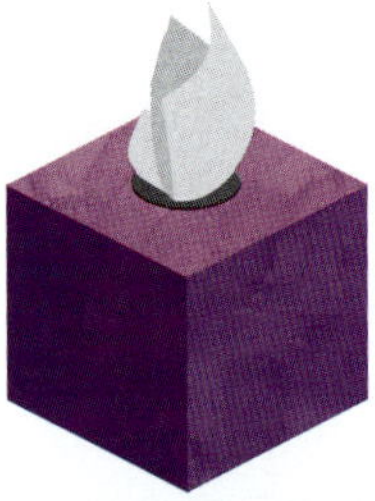		

Complete Worksheet **3**: pages **71–72**

Grouping 3D Shapes

Lesson 4

Explore

The children are taking turns to group the items based on their shape.

How did Jacob group the items?

Master

1

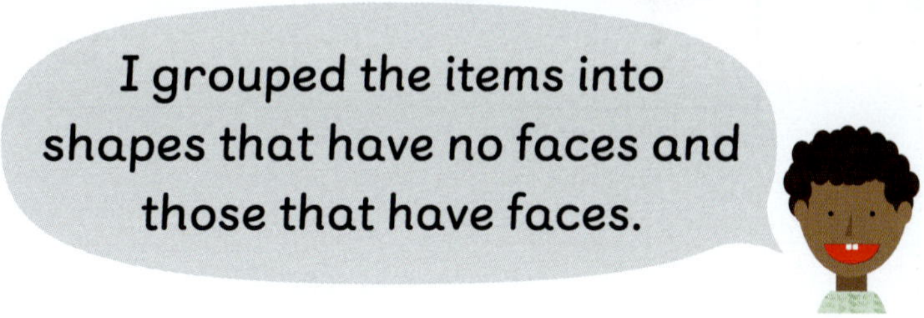

No faces	Have faces
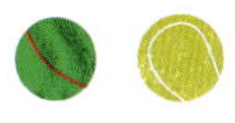	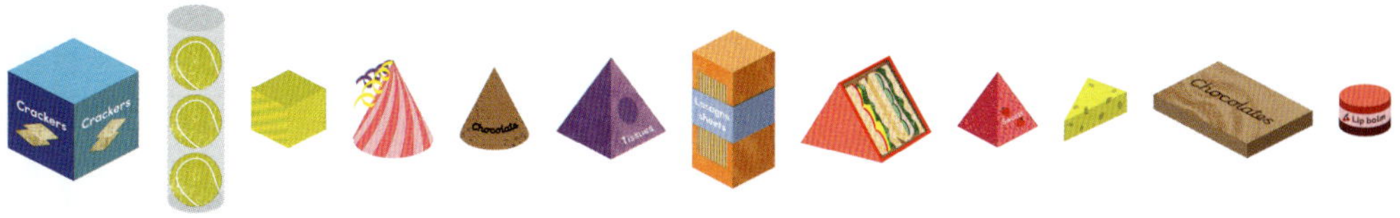

Spheres have no faces.
All of the other shapes have at least one face.

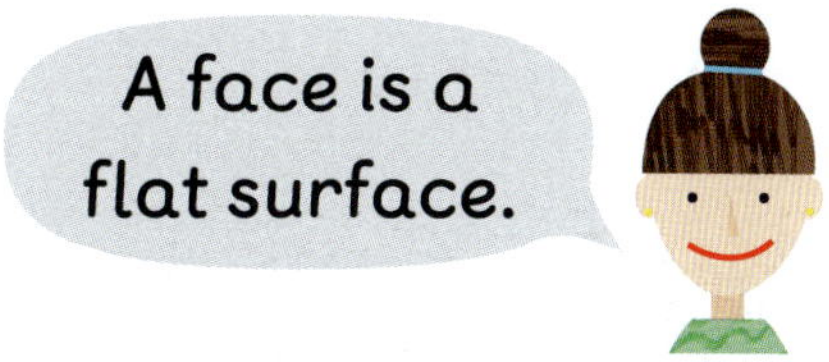

2

I grouped them into shapes that have curved surfaces and those that have only faces.

Curved surfaces	No curved surfaces
	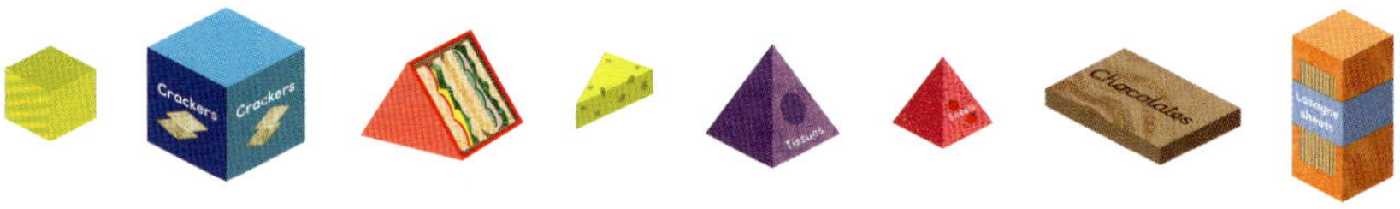

3

I grouped them by the shape of their faces.

Shapes with triangular faces	Shapes with no triangular faces	Shapes with no faces

Are there other ways to group these shapes?

Activity Time

Group work (4 pupils)

You will need:

1. Give your partner 4 shapes.
2. Put the remaining shapes into groups.
3. Ask your partner to put the 4 shapes into the groups you have made.
4. If your partner does not put them in the right group, give them a clue about how you grouped the shapes.
5. Talk about how the shapes were grouped.
6. Repeat the activity, taking turns.

Guided Practice

Which is the odd one out in each group?
Explain your answers.

1

2

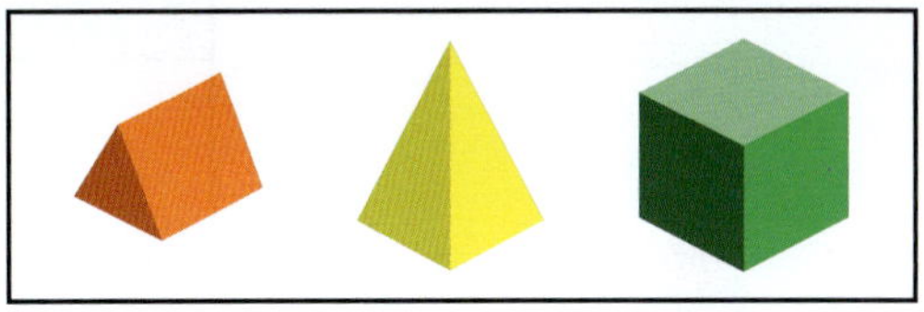

3

4

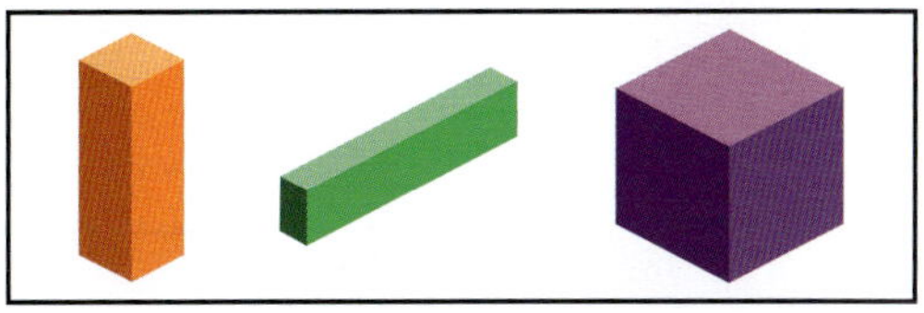

Complete Worksheet 4: pages **73–74**

Forming 3D Structures

Lesson 5

Explore

Charles made these structures.

What shapes does Holly need to make the same structures?

Master

1 What shapes did Charles use to make this structure?

	Name	Number of shapes used
	cone	2
	cuboid	2
	prism	0
	cube	4
	cylinder	2

2 What shapes did Charles use to make this structure?

	Name	Number of shapes used
	cone	
	cuboid	
	prism	
	cube	
	cylinder	

Activity Time

Partner work

You will need:

1. You and your partner agree on the number of shapes you will each use to make a structure.
2. Make your structures using the shapes.
3. Talk about what is the same and what is different about each of your structures.
4. Repeat the activity using a different number of shapes.

Guided Practice

How many shapes are used in each structure?

1

	Name	Number of shapes used
	cone	
	cuboid	
	prism	
	cube	
	cylinder	

2

Name	Number of shapes used
cone	
cuboid	
prism	
cube	
cylinder	

Complete Worksheet **5**: pages **75–76**

How can Ravi and Lulu share the piece of paper equally?

Chapter 13
Fractions

Showing Equal Parts

Lesson 1

Explore

Ravi, Lulu, Amira and Elliott want to share a piece of paper equally. They draw different possible ways to do this.

Which ones show the paper cut in equal parts?

Master

1

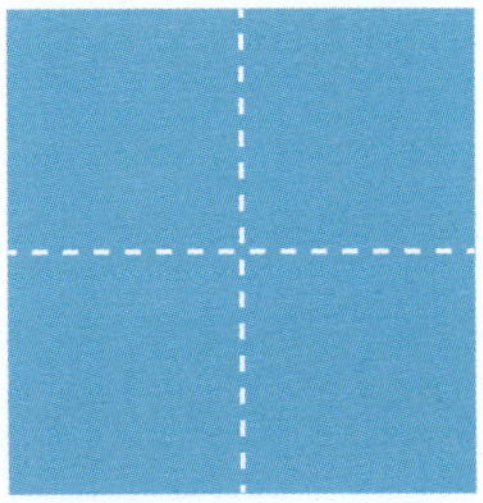

This piece of paper is cut in equal parts.

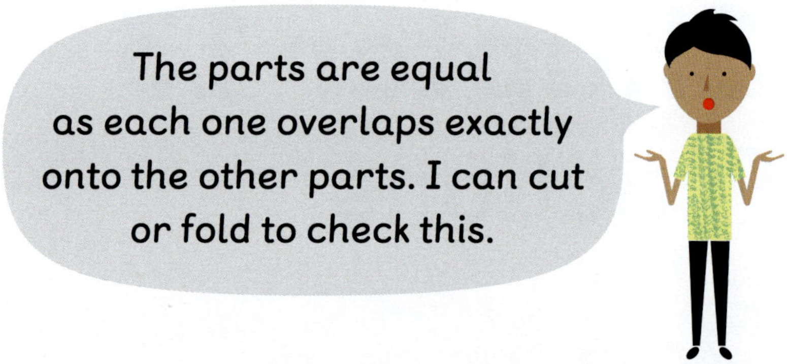

2 These pieces of paper show equal parts too.

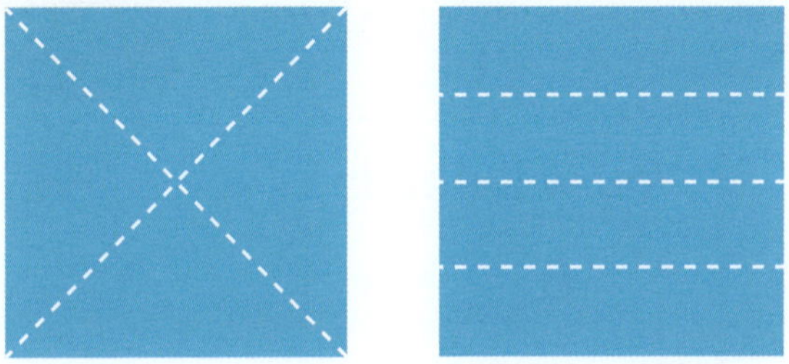

3 This piece of paper is not cut in equal parts.
Each part does not overlap exactly onto all the other parts.

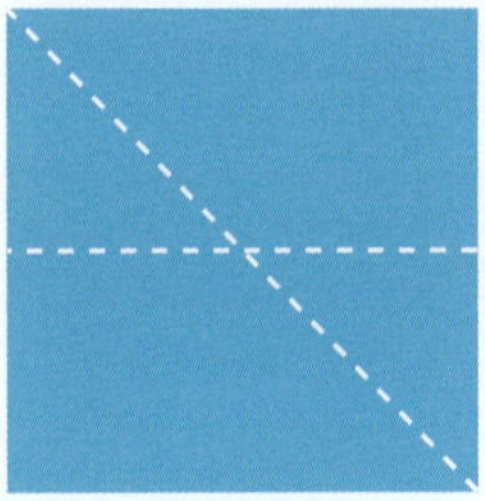

4 Which of these show equal parts?

A B C

Activity Time

Partner work

You will need:
6 

1. Each partner takes 1 piece of paper. Show your partner one way of making 2 equal parts.
2. Have your partner show you a different way using their piece of paper.
3. Repeat this activity to show how you can make 4 equal parts and then 3 equal parts in different ways with new pieces of paper.

Guided Practice

1 Which shapes show equal parts?
Talk to your partner about how you know they are equal.

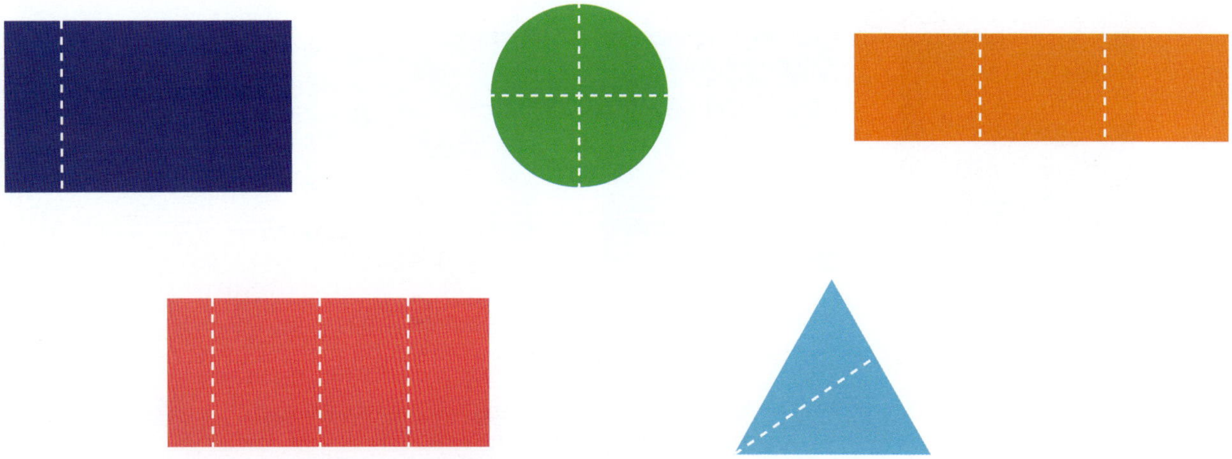

Complete Worksheet **1**: pages **91–92**

Showing Half and Quarter

Lesson 2

Explore

Elliott needs to cut the blue paper into 2 equal parts.
He needs to cut the orange paper into 4 equal parts.
How can he do this?

Master

1 Fold the paper once so that two parts overlap onto each other exactly. Then unfold the paper.

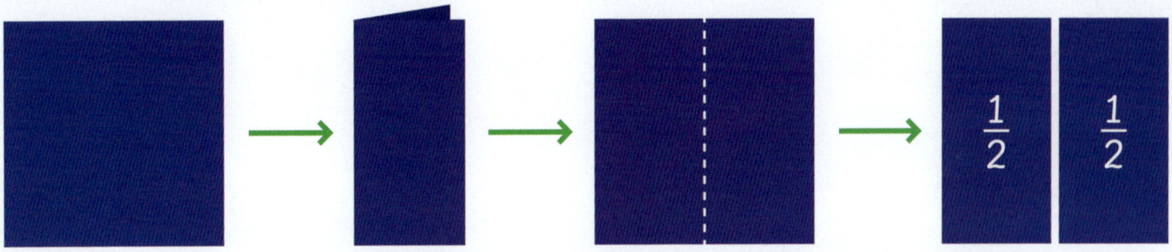

Cut along the fold line to get 2 equal parts.

Each part is 1 out of 2 equal parts.

We write '1 out of 2' as $\frac{1}{2}$.

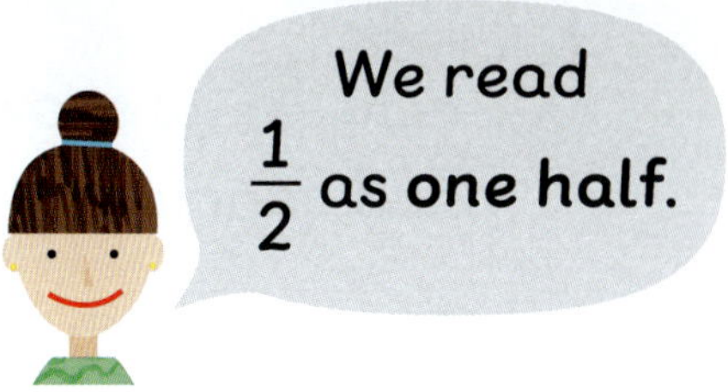

is $\frac{1}{2}$ or one half of the whole.

Two halves make a whole.

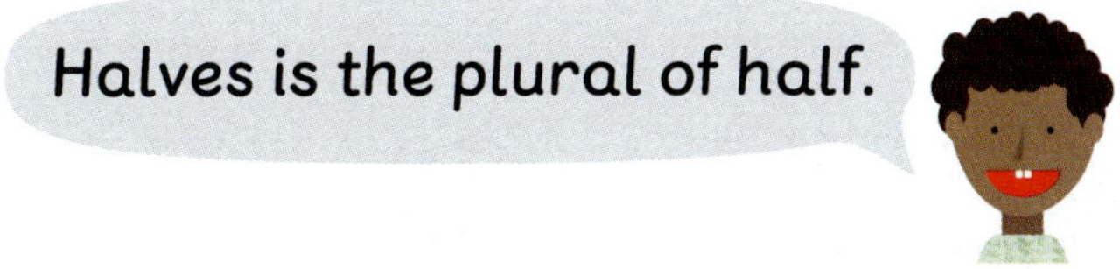

$\frac{2}{2}$ is a whole.

2 Fold the paper twice so that the parts overlap onto each other exactly. Then, unfold the paper.

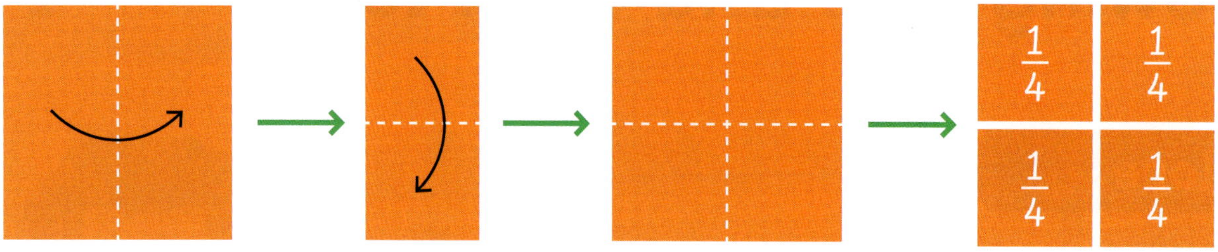

Cut along the fold lines to get 4 equal parts.

Each part is 1 out of 4 equal parts.

We write '1 out of 4' as $\frac{1}{4}$.

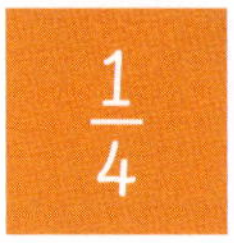

is one quarter of the whole.

Four quarters make a whole.

$\frac{4}{4}$ is a whole.

Activity Time

Partner work

You will need:

(1) Fold  to show halves and quarters.

(2) Together with your partner, make a poster with all the different ways you can find to show halves and quarters.

(3) Talk to your partner about how you can use what you know about halves to show quarters.

Guided Practice

1 How much of each shape is shaded? Use $\frac{1}{2}$ or $\frac{1}{4}$ to fill in the blanks.

(a)

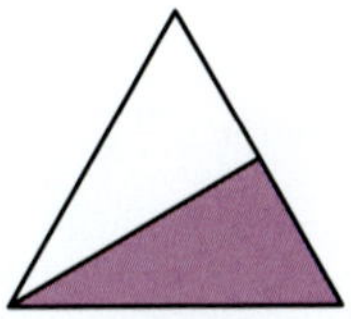

☐

(b)

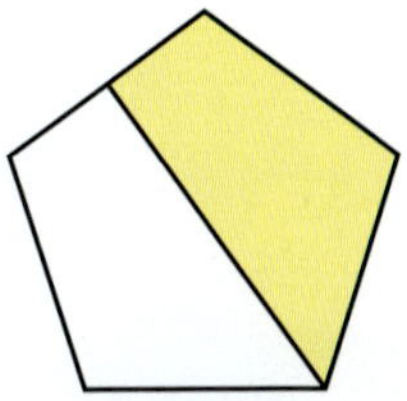

☐

(c)

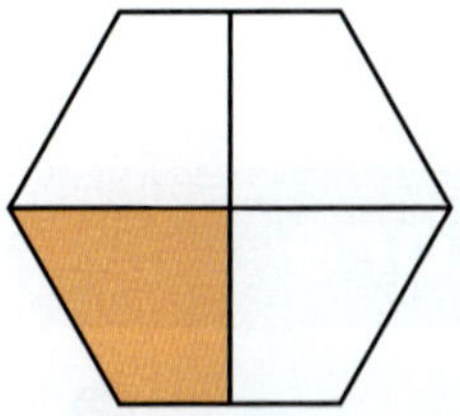

☐

2 Find the shapes that show $\frac{1}{2}$ or $\frac{1}{4}$ shaded.

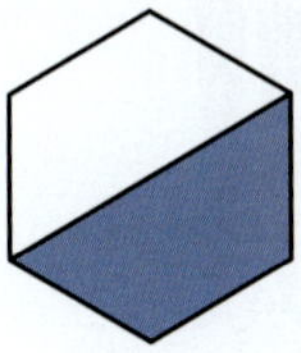

A

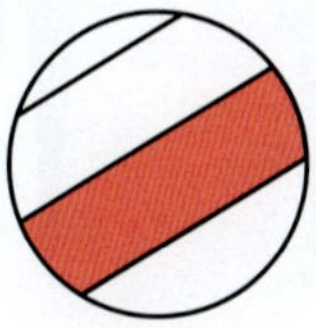

B

C

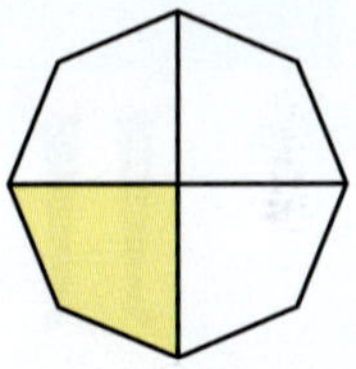

D

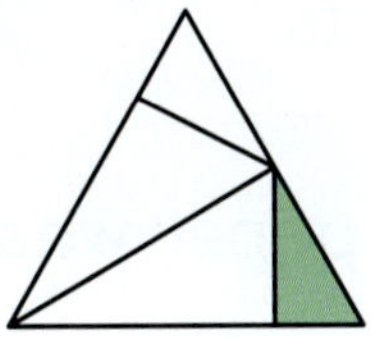

E

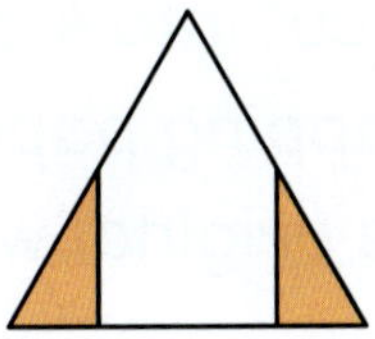

F

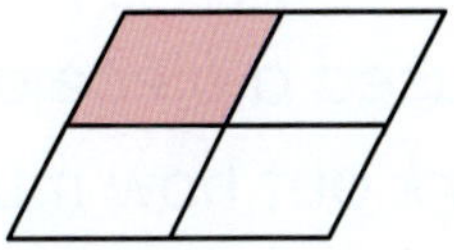

G

3 Use $\frac{1}{2}$ or $\frac{1}{4}$ to fill in the blanks.

(a) We write one quarter as ☐.

(b) We write one half as ☐.

(c) We write one fourth as ☐.

Complete Worksheet **2**: pages **93–96**

Showing Quarters

Lesson 3

Explore

A large piece of wrapping paper was cut into 4 equal parts.
3 friends each used a piece of the wrapping paper.
How can we work out how much of the original wrapping paper was used?

Master

1 What fraction of the wrapping paper was used?

The wrapping paper was cut into 4 equal parts.
The name of each part is a quarter or a fourth.
3 of the 4 equal parts of the wrapping paper were used.

$\frac{3}{4}$ of the wrapping paper was used.

2 Show quarters by shading the shapes.

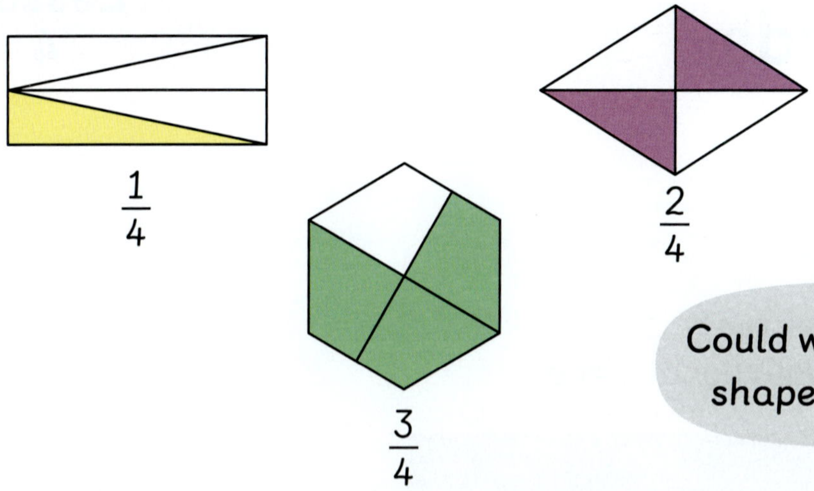

Guided Practice

Which of these show:

1 $\frac{1}{4}$ of the square or rectangle shaded?

2 $\frac{3}{4}$ of the square or rectangle shaded?

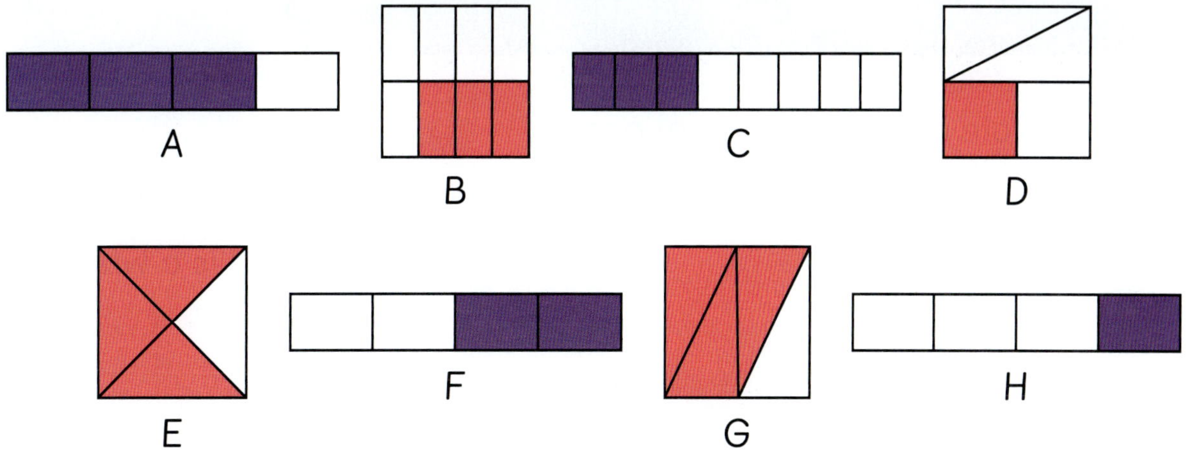

Complete Worksheet **3**: pages **97–98**

Showing Thirds

Lesson 4

Explore

Ravi's mum cuts the wood into equal pieces.

She uses 2 of the 3 equal pieces.
How much of the wood did she use?

Master

1

The whole is cut into 3 equal parts.

Each part is $\frac{1}{3}$ of the whole.

We write one third as $\frac{1}{3}$.
$\frac{3}{3}$ or 3 thirds is a whole.

2

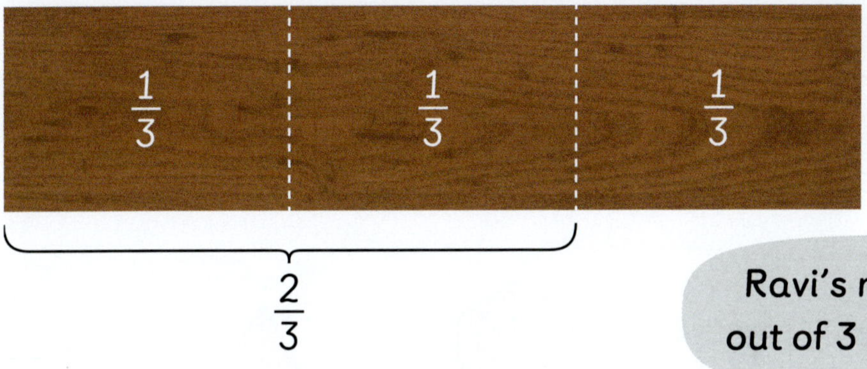

Ravi's mum used $\frac{2}{3}$ or 2 thirds of the wood.

3

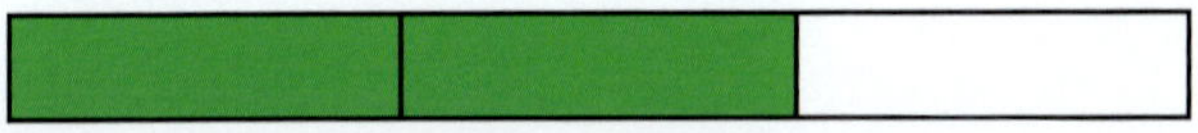

$2 \longrightarrow$ numerator
$3 \longrightarrow$ denominator

$\frac{2}{3}$ is a fraction.

Its numerator is 2 and its denominator is 3.

The **numerator** in a fraction shows the number of equal parts.

The **denominator** in a fraction shows the number of equal parts the whole is divided into.

4

$\frac{1}{2}$

one half

$\frac{1}{4}$

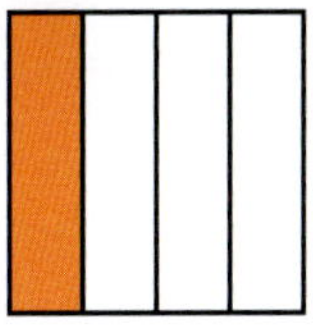

one quarter

$\frac{1}{3}$

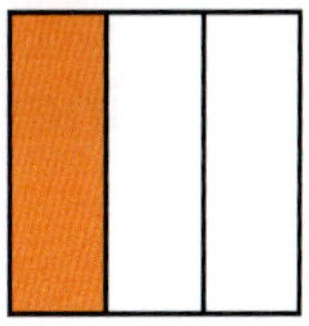

one third

Fractions that have a numerator of 1 are called **unit fractions**.

$\frac{1}{2}$, $\frac{1}{4}$ and $\frac{1}{3}$ are unit fractions.

Guided Practice

1 How much of the shape is shaded?
Match.

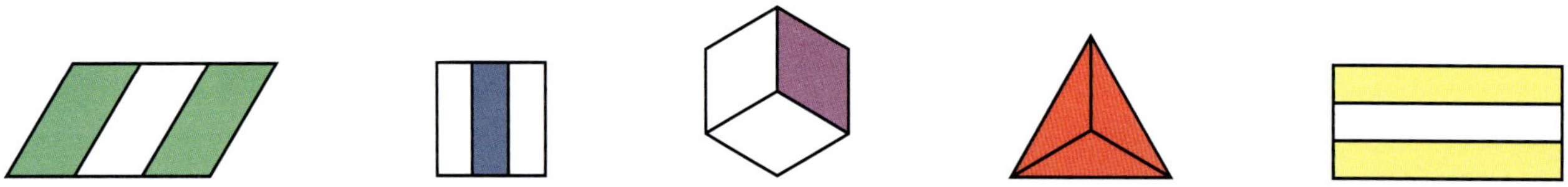

$\frac{1}{3}$ $\frac{2}{3}$ $\frac{3}{3}$

2 Use $\frac{1}{3}$, $\frac{2}{3}$ or $\frac{3}{3}$ to describe how full the container is.

(a)

The container is ☐ full.

(b)

The container is ☐ full.

(c)

The container is ☐ full.

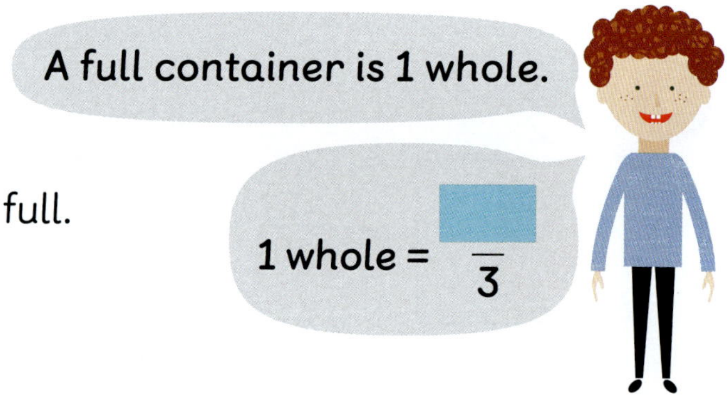

Complete Worksheet **4**: pages **99–100**

Naming Fractions

Lesson 5

Explore

How has Elliott cut his pancake?

Master

1

This is a whole.

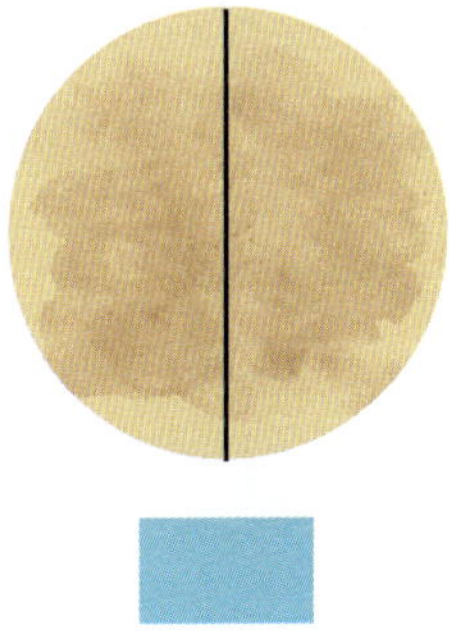

$\frac{}{2}$

The whole is divided into 2 equal parts.

Each part is a half or $\frac{1}{2}$.

When the whole is divided into 2 equal parts, the denominator of the fraction is 2.

Elliott has cut his pancake into halves.

2 How has Lulu cut her pancake?

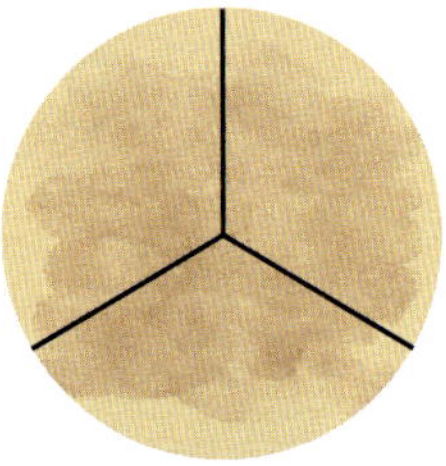

$\frac{\square}{3}$

The whole is divided into 3 equal parts.

Each part is a third or $\frac{1}{3}$.

When the whole is divided into 3 equal parts, the denominator of the fraction is 3.

Lulu has cut her pancakes into thirds.

3 How has Ravi cut his pancake?

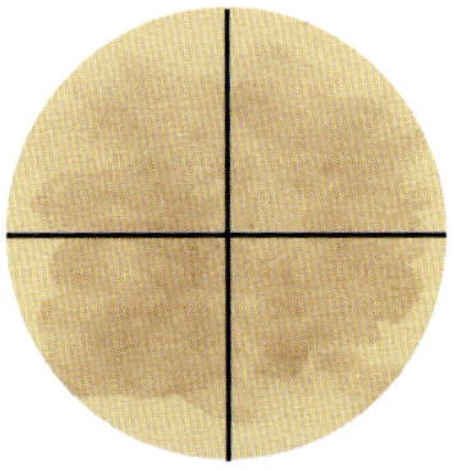

$\frac{\square}{4}$

The whole is divided into 4 equal parts.

Each part is a quarter or $\frac{1}{4}$.

When the whole is divided into 4 equal parts, the denominator of the fraction is 4.

Ravi has cut his pancakes into quarters.

4 The whole has been divided into halves.

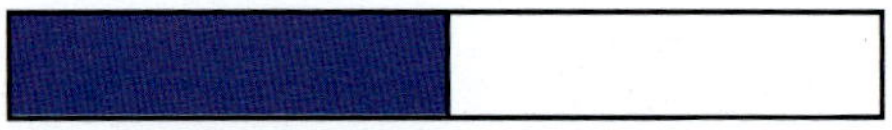

One half is written as $\frac{1}{2}$.

The numerator is 1.

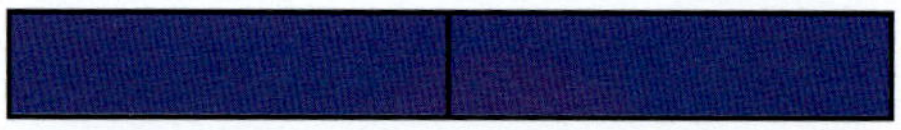

Two halves is written as $\frac{2}{2}$.

The numerator is 2.

5 The whole has been divided into thirds.

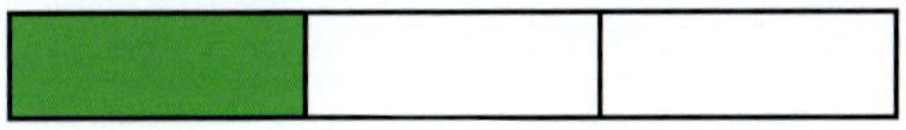

One third is written as $\frac{1}{3}$.

The numerator is 1.

Two thirds is written as $\frac{2}{3}$.

The numerator is 2.

6 The whole has been divided into quarters or fourths.

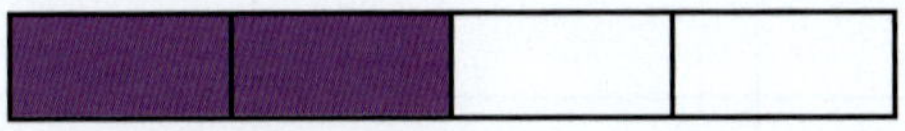

Two quarters is written as $\frac{2}{4}$.

The numerator is ______.

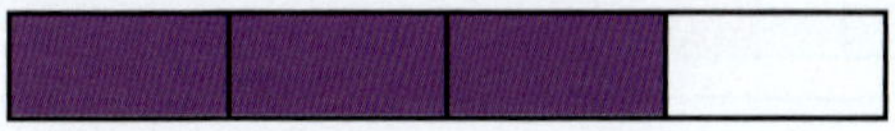

Three quarters is written as $\frac{3}{4}$.

The numerator is ______.

7

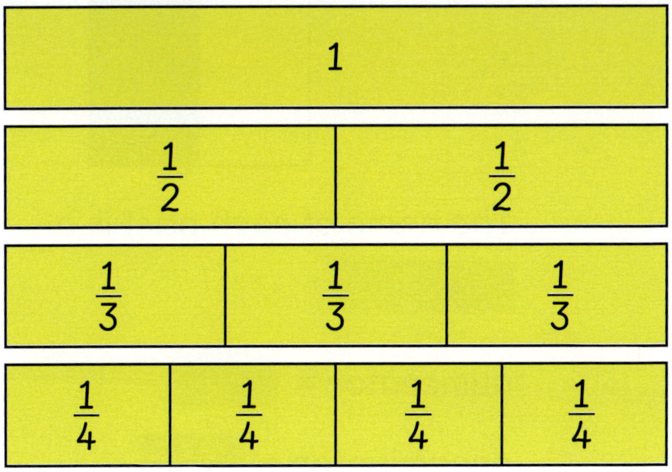

Guided Practice

1 Complete the table.

Picture	Symbol	Word	Numerator	Denominator
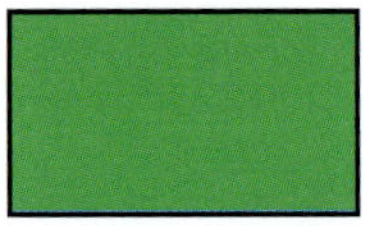	1	one		
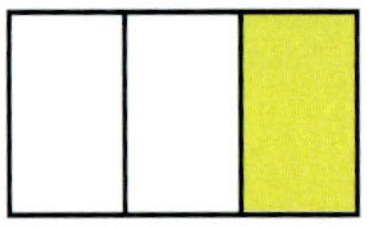	$\frac{1}{3}$	one third	1	3
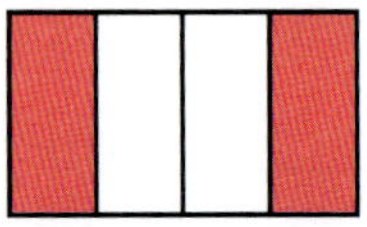				
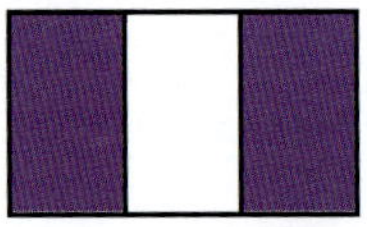				
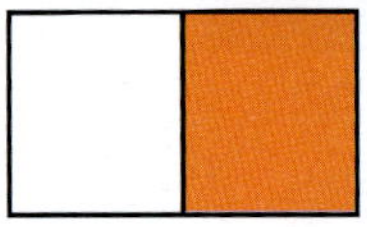				

2 What fraction of the square is shaded?

The name of each part is a ______.

numerator = ______

denominator = ______

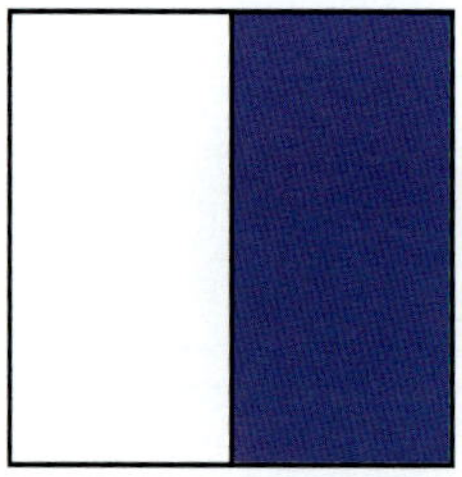

The name of each part is a ______.

numerator = ______

denominator = ______

Complete Worksheet **5**: pages **101–104**

Making a Whole

Lesson 6

Explore

A baker cuts a large brownie into 3 equal pieces.
He sells 1 piece of brownie in the morning.
He sells the remaining pieces of brownie in the afternoon.

What fraction of the whole brownie did he sell in the morning?
What fraction of the whole brownie did he sell in the afternoon?

Master

1

Each piece is 1 third.
3 thirds make 1 whole.

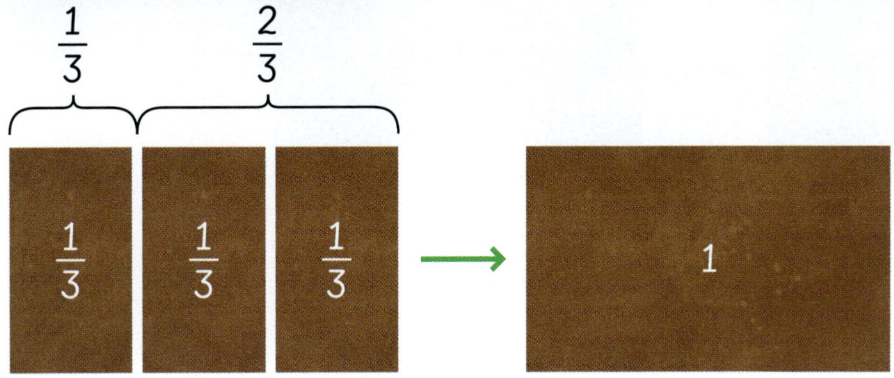

$\frac{1}{3}$ and $\frac{2}{3}$ make 1 whole.

The baker sold $\frac{1}{3}$ of the brownie in the morning.

He sold $\frac{2}{3}$ of the brownie in the afternoon.

2 What if the brownie was cut into 4 equal pieces?

Each piece is 1 quarter.
4 quarters make 1 whole.

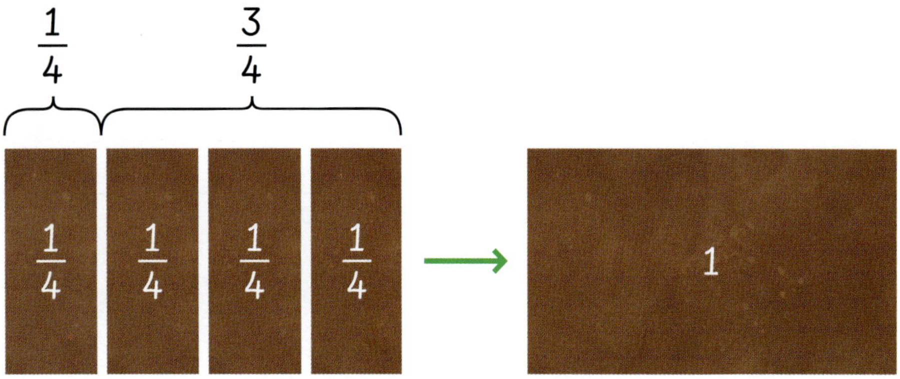

$\frac{1}{4}$ and $\frac{3}{4}$ make 1 whole.

3

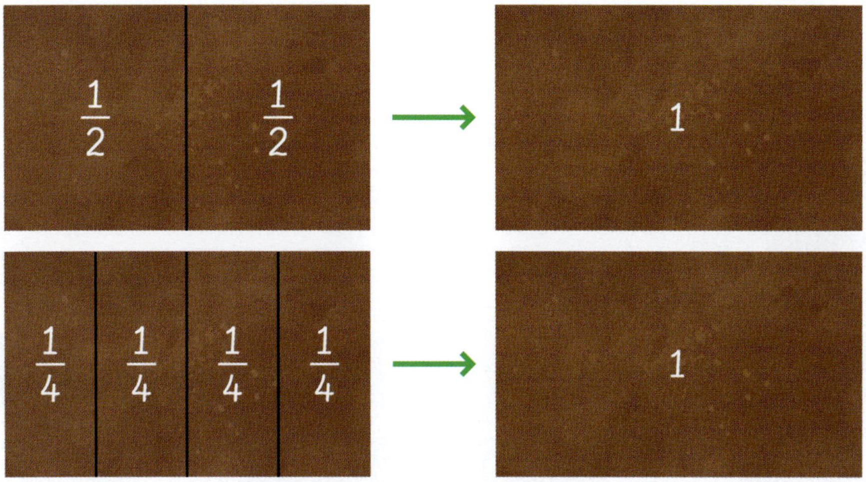

$\frac{1}{4}$ and $\frac{1}{4}$ make $\frac{1}{2}$.

$\frac{2}{4} = \frac{\square}{2}$

$\frac{\square}{4}$ and $\frac{\square}{4}$ make 1.

Guided Practice

1 The shaded parts show a fraction.
Match the fractions to make 1 whole.

2 Fill in the blanks.

(a)

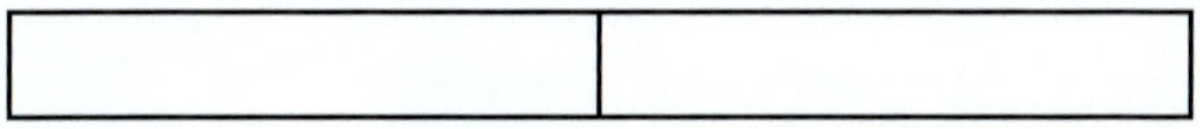

$\frac{\_\_\_}{2}$ and $\frac{\_\_\_}{2}$ make 1 whole.

(b)

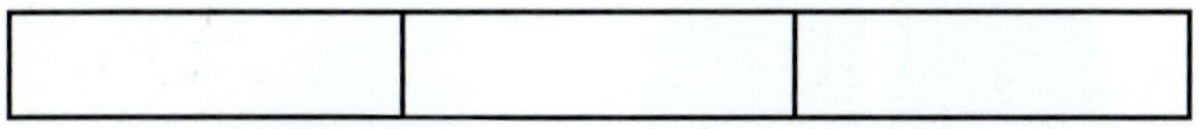

$\frac{1}{3}$ and ___ make 1 whole.

(c)

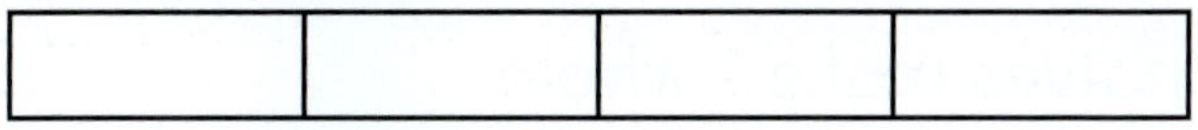

$\frac{1}{4}$ and ___ make 1 whole.

3 Fill in the blanks.

(a) ___ and $\frac{1}{3}$ make 1 whole.

(b) $\frac{2}{4}$ and ___ make 1 whole.

(c) $\frac{3}{4}$ and ___ make 1 whole.

(d) $\frac{1}{4}$ and ___ is the same as $\frac{1}{2}$.

Complete Worksheet **6**: pages **105–106**

Counting in Halves

Lesson 7

Explore

A baker cuts his cakes in half and sells them.

How many cakes does he have left to sell altogether?

Master

1

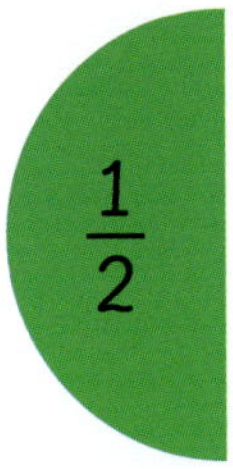

This is 1 half.

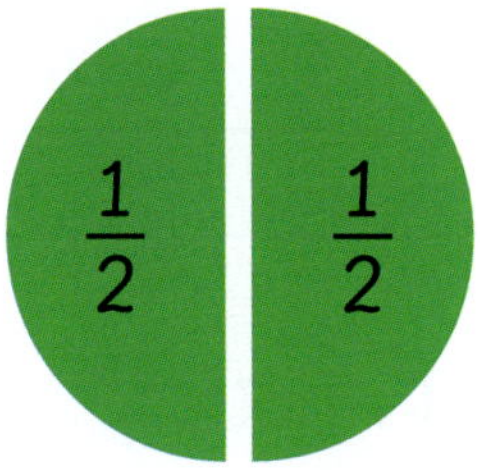

These are 2 halves.
2 halves make 1 whole.

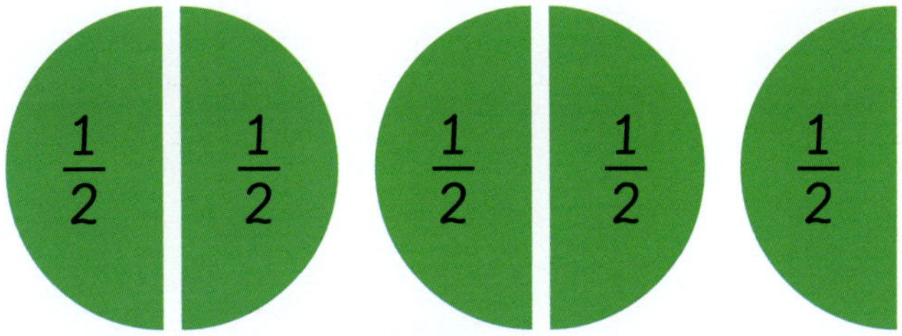

These are 5 halves.
5 halves make 2 wholes and a half.

This is 2 and $\frac{1}{2}$.

We write 2 and $\frac{1}{2}$ as $2\frac{1}{2}$.

The baker has $2\frac{1}{2}$ cakes left to sell altogether.

2 Count on in halves.

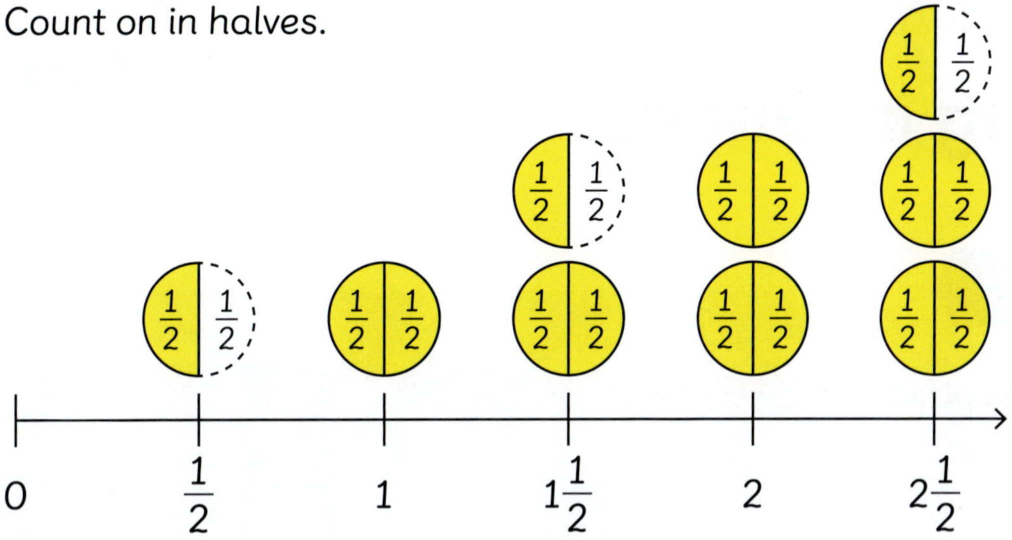

Guided Practice

1 What number is shown by the shaded parts?

(a)

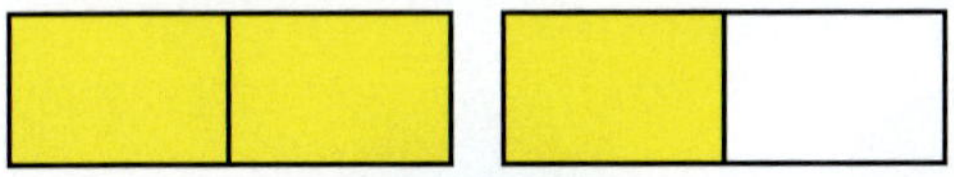

(b)

2 Fill in the missing numbers.

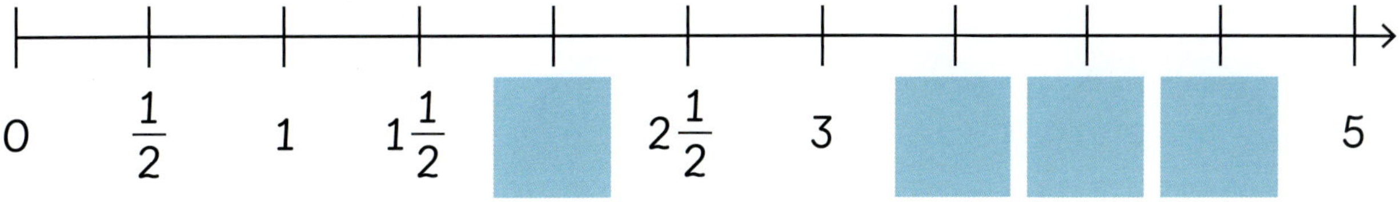

3 Fill in the missing numbers by counting in halves.

$6, 6\frac{1}{2}, 7,$ ___ , ___ $, 8\frac{1}{2}$

Complete Worksheet 7: pages 107–108

Counting in Quarters

Lesson 8

Explore

Amira has some pizza left over after her party.
How many pizzas are left over?

Master

1

This is $\frac{1}{4}$.

These are 4 quarters.
4 quarters make 1 whole.

There are 3 whole pizzas and 1 quarter of a pizza.

There are 3 and $\frac{1}{4}$ pizzas.

We write 3 and $\frac{1}{4}$ as $3\frac{1}{4}$.

There are $3\frac{1}{4}$ pizzas left over.

2 Count on in quarters.

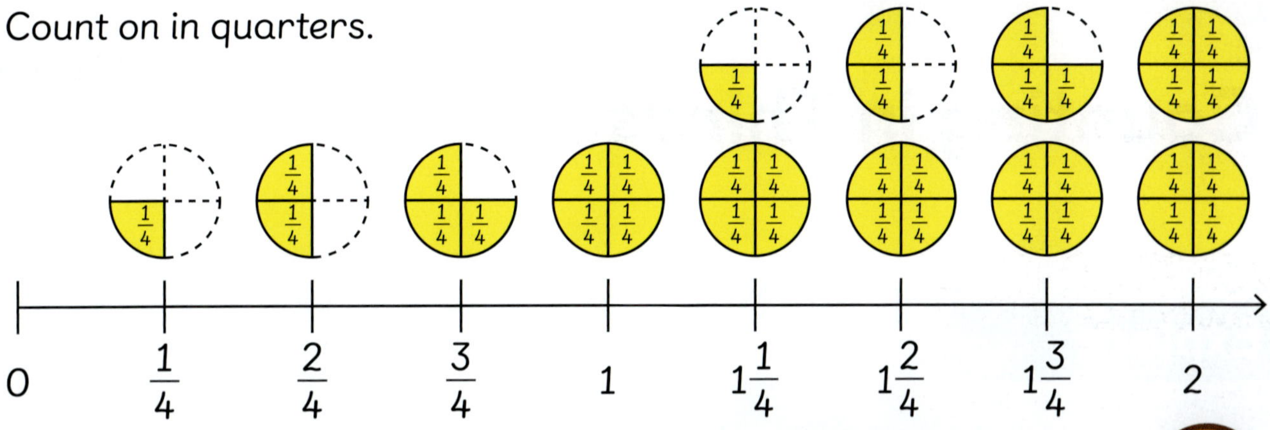

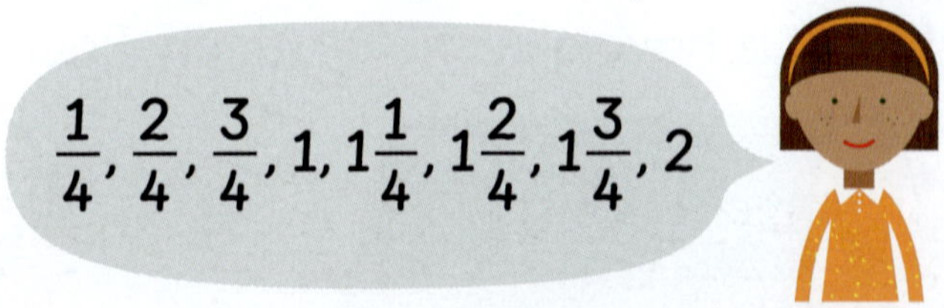

Guided Practice

1 What number is shown by the shaded parts?

(a)

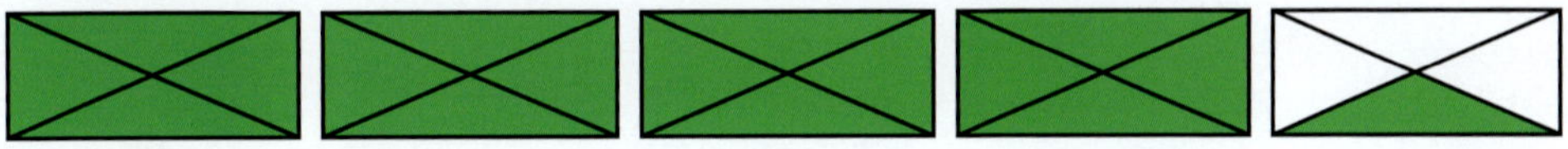

(b)

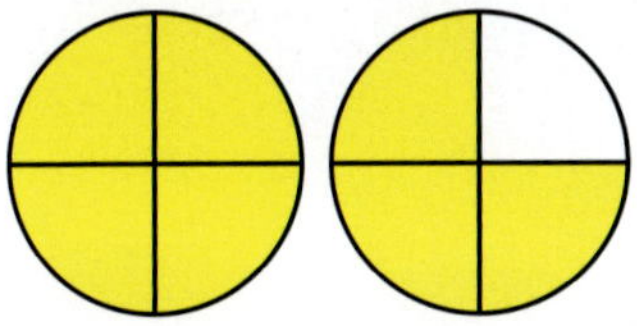

2 Fill in the missing numbers.

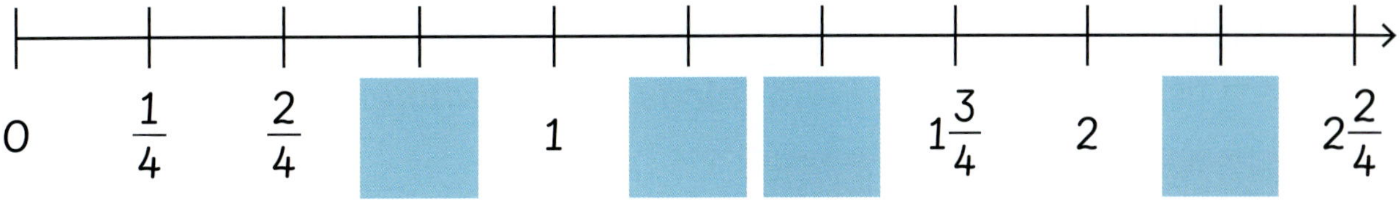

3 Fill in the missing numbers by counting in quarters.

$8\frac{2}{4}$, $8\frac{3}{4}$, 9, ______, ______, $9\frac{3}{4}$

Complete Worksheet 8: pages 109–110

Counting in Thirds

Lesson 9

Explore

Elliott is having sushi for lunch. Each sushi roll has been cut into 3 pieces.

How much sushi does he have left?

Master

1

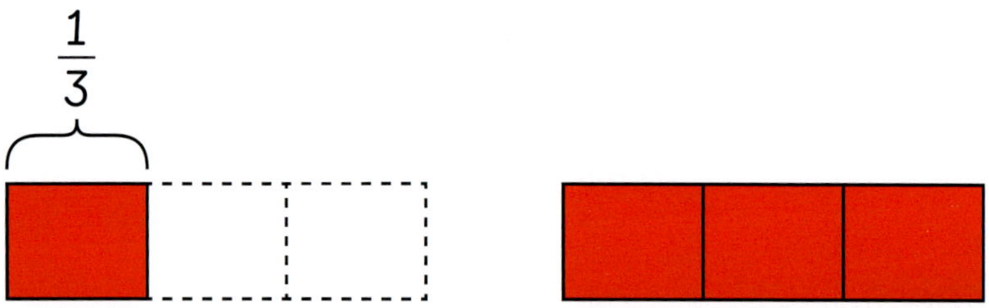

This is $\frac{1}{3}$.

These are 3 thirds.
3 thirds make 1 whole.

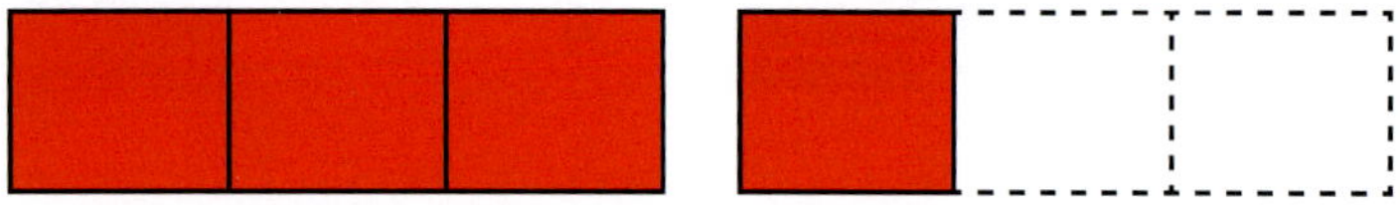

There is 1 whole roll of sushi and $\frac{1}{3}$ roll of sushi.

Elliott has $1\frac{1}{3}$ rolls of sushi.

2 Count in thirds.

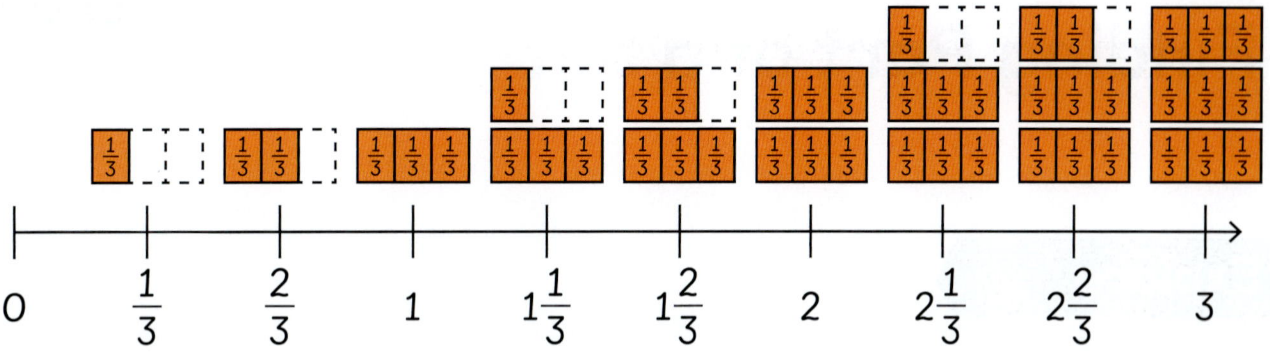

Guided Practice

1 What number is shown by the shaded parts?

(a)

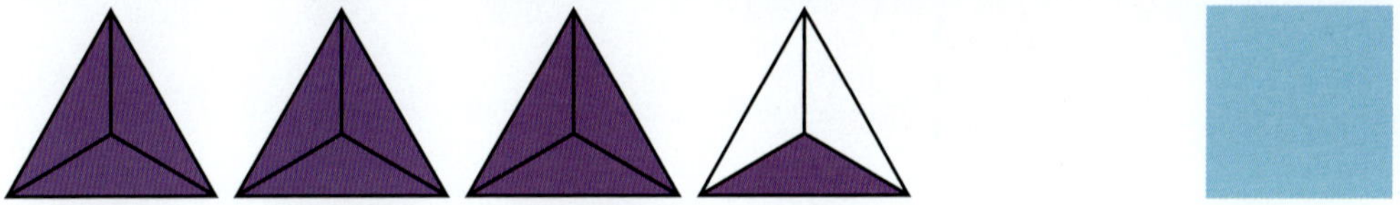

(b)

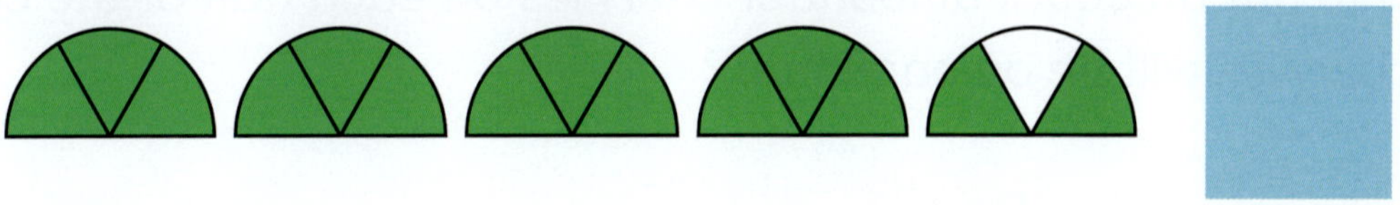

2 Fill in the missing numbers.

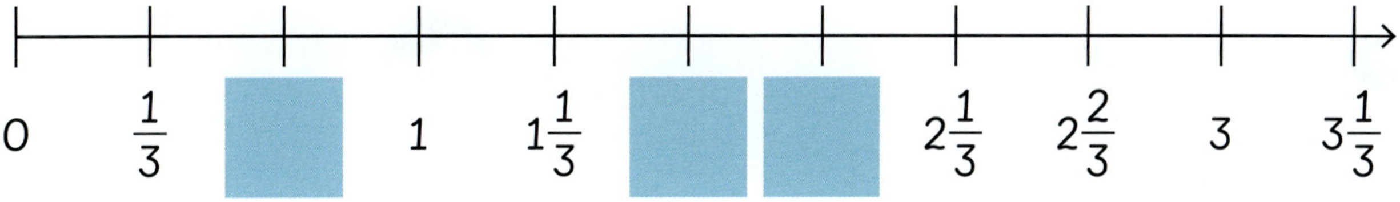

3 Fill in the missing numbers by counting in thirds.

$8\frac{1}{3}$, $8\frac{2}{3}$, 9, $9\frac{1}{3}$, ___, ___

Complete Worksheet **9**: pages **111–112**

Finding Part of a Set

Lesson
10

Explore

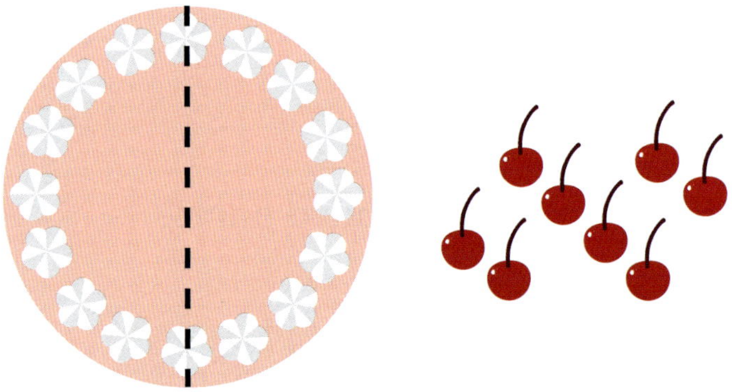

Amira wants to put an equal amount of cherries on each half of the cake. How many cherries will be on each half?

Master

1

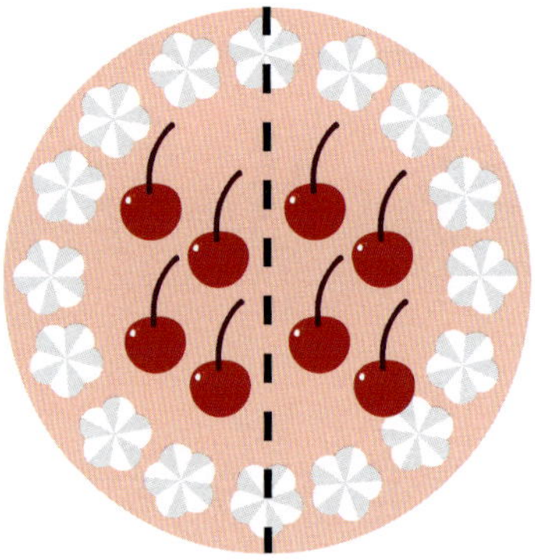

There will be 4 cherries on each half of the cake.

2 What is $\frac{1}{2}$ of 6 crayons?

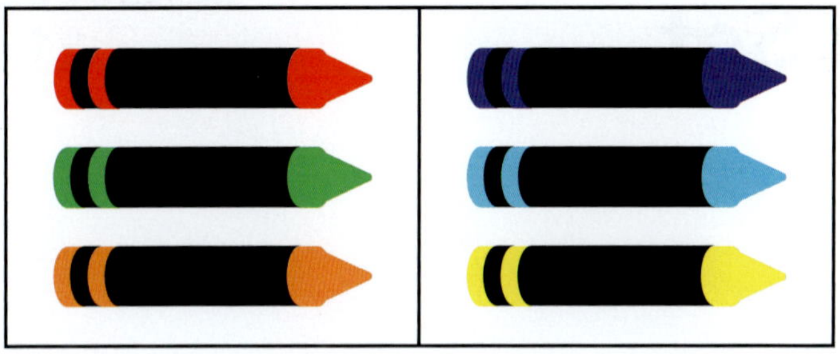

Guided Practice

1 Find $\frac{1}{2}$ of each group.

(a)

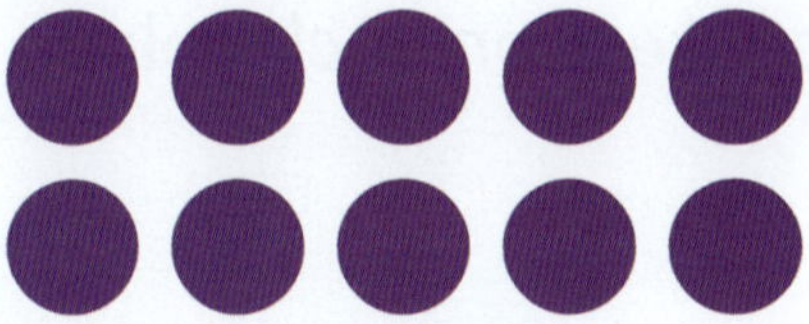

$\frac{1}{2}$ of 10 =

(b)

$\frac{1}{2}$ of 18 =

2 Fill in the blanks.

(a) $\frac{1}{2}$ of 12 =

(b) $\frac{1}{2}$ of 16 =

(c) $\frac{1}{2}$ of 20 =

Complete Worksheet **10**: pages **113–114**

Finding Part of a Set

Lesson 11

Explore

How can Charles put an equal number of his stickers on each colour on his folder?

Master

1

Charles can put 4 stickers on each part.

2 What is $\frac{1}{3}$ of 6 stickers?

3 What is $\frac{1}{3}$ of 9 stickers?

$\frac{1}{3}$ of 9 is 3.

Guided Practice

1 Find $\frac{1}{3}$ of each group.

(a)

$\frac{1}{3}$ of 12 = ___

(b)

$\frac{1}{3}$ of 18 = ___

2 Fill in the blanks.

(a) $\frac{1}{3}$ of 15 = ___

(b) $\frac{1}{3}$ of 27 = ___

Complete Worksheet **11**: pages **115–116**

Finding Part of a Set

Lesson 12

Explore

Is Elliott correct?

Master

1

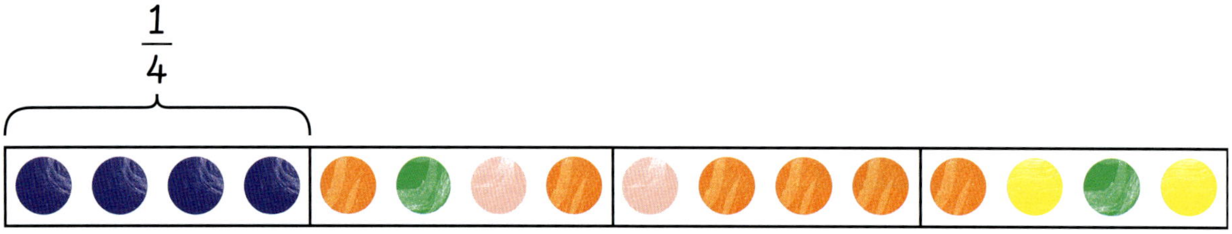

16 marbles are shared equally into 4 groups.

Each group is $\frac{1}{4}$ of the total number of marbles.

$\frac{1}{4}$ of 16 is 4.

Elliott is correct.

2 Find $\frac{1}{4}$ of 12.

Each group is $\frac{1}{4}$ of 12.

$\frac{1}{4}$ of 12 = 3.

3 What is $\frac{1}{4}$ of 8?

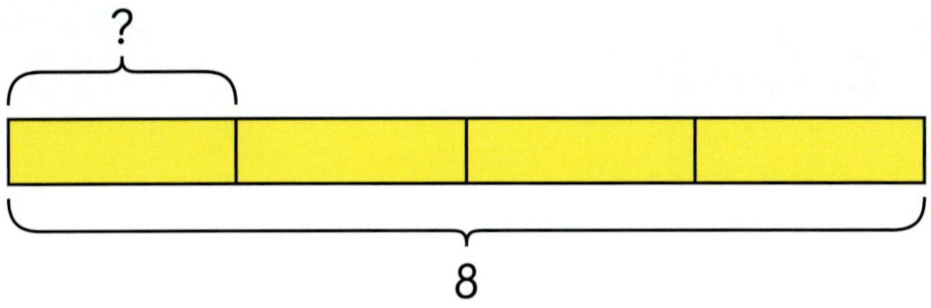

$8 \div 4 =$

$\frac{1}{4}$ of $8 =$

Guided Practice

1

$\frac{1}{4}$ of $20 =$

2

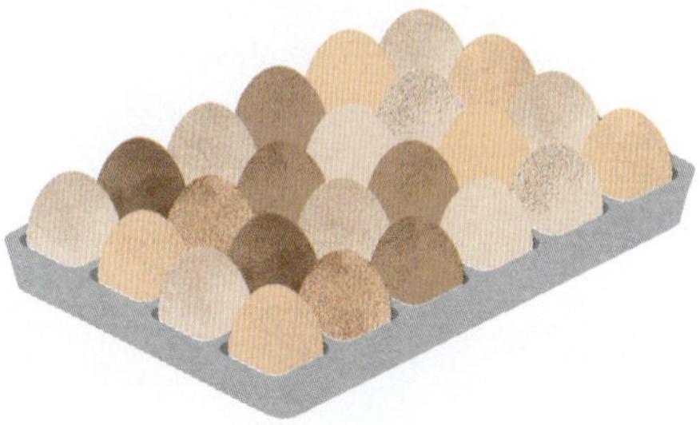

$\frac{1}{4}$ of $24 =$

3

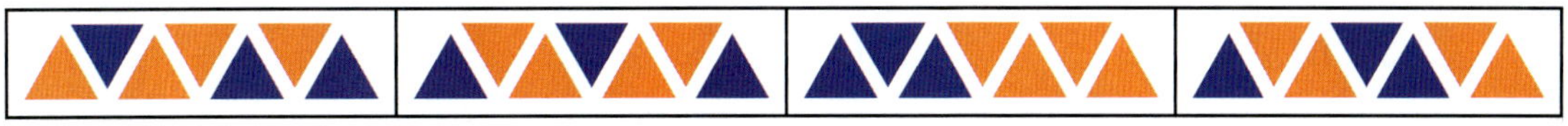

(a) $\frac{1}{4}$ of $28 =$ (b) $\frac{2}{4}$ of $28 =$ (c) $\frac{3}{4}$ of $28 =$

4 (a) $\frac{1}{4}$ of $40 =$ (b) $\frac{2}{4}$ of $40 =$ (c) $\frac{3}{4}$ of $40 =$

Complete Worksheet **12**: pages **117–118**

Finding Part of a Quantity

Lesson 13

Explore

Amira has 20 cm of tape.

She uses $\frac{1}{2}$ of the tape on each box.

What is the length of tape she used on each box?

Master

1 Find $\frac{1}{2}$ of 20 cm.

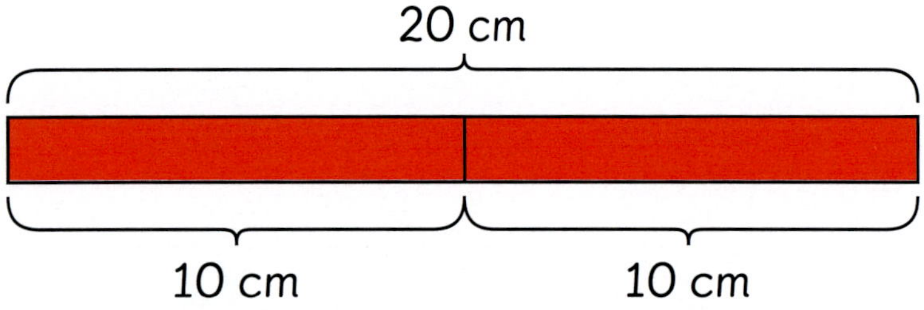

The tape can be cut into 2 parts of equal length.

$\frac{1}{2}$ of 20 cm = 10 cm

2 Find $\frac{1}{4}$ of 8 cm.

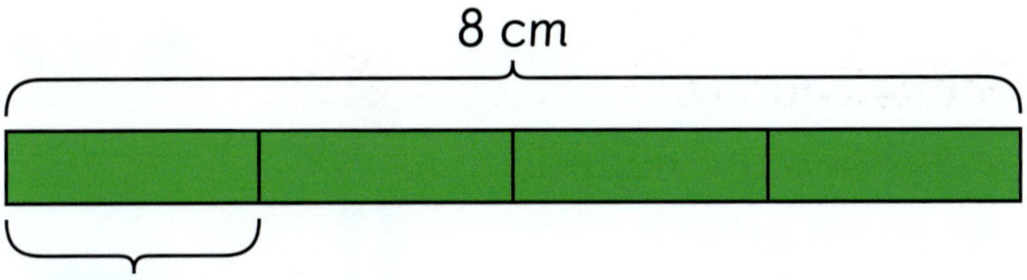

____ cm

There are ____ parts of equal length.

$\frac{1}{4}$ of 8 cm = ____ cm

Guided Practice

1

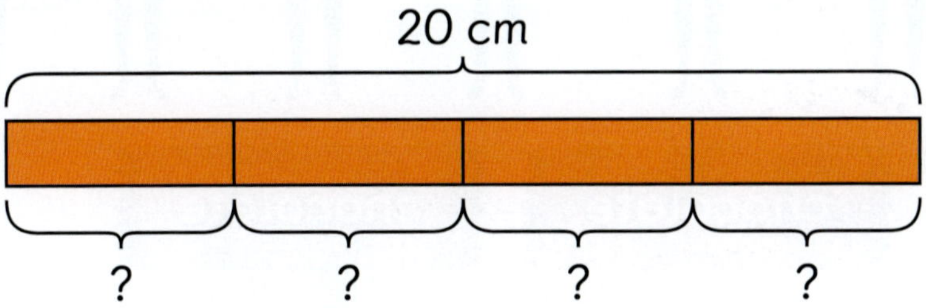

There are ____ parts of equal length.

$\frac{1}{4}$ of 20 cm = ____ cm

2

$\frac{1}{3}$ of 18 cm = ____ cm

3 (a) $\frac{1}{2}$ of 18 cm = ____ cm (b) $\frac{1}{4}$ of 12 cm = ____ cm

Complete Worksheet **13**: pages **119–120**

Mind Challenge

9 children share 3 bars of chocolate equally.
6 adults share 2 bars of the same chocolate equally.
Who gets more chocolate, a child or an adult?

My Maths Journal

Think of fractions you see around you.
Record them in your journal.

For example:

$\frac{3}{4}$ of the sandwich is left.

Review

I can...

- ☐ show equal parts.
- ☐ make and show halves, quarters and thirds.
- ☐ name and write a fraction.
- ☐ make one whole with two fractions.
- ☐ count wholes with halves, quarters and thirds.
- ☐ find part of a set and a quantity.

How long have Ravi and Lulu been climbing for?

Chapter 14
Time

Telling and Writing Time to 5 Minutes

Lesson 1

Explore

Jacob timed himself finishing a problem.

start

finish

At what time did he finish the problem?

Master

1

1 minute

Each mark on a clock shows 1 minute.

5 minutes

It takes 5 minutes for the minute hand to move from 12 to 1.

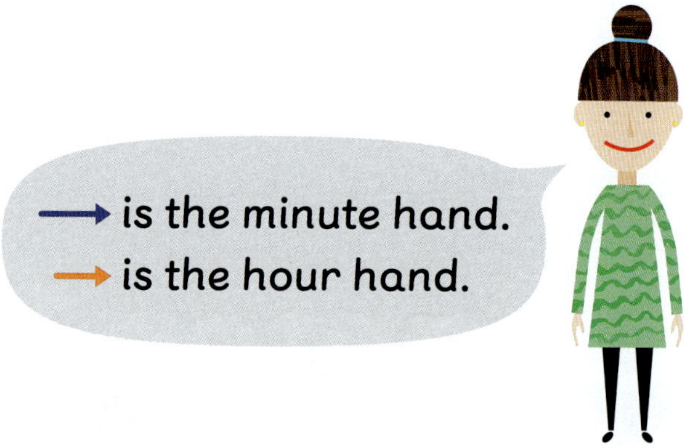

2

It takes 10 minutes for the minute hand to move from 12 to 2.
The time is 10 minutes after 9 o'clock.
We say it is 10 minutes past 9.
Jacob finished the problem at 9:10.

3

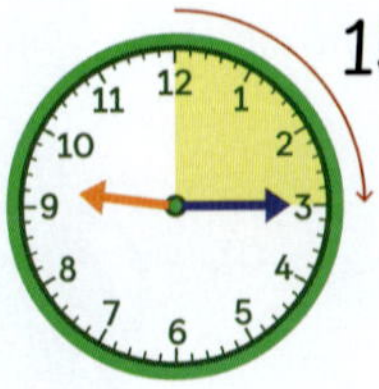

The time is 9:15.
We can say that 9:15
is quarter past 9.

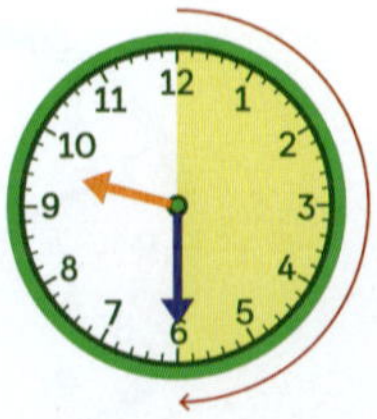

The time is 9:30.
We can say that 9:30
is half past 9.

Guided Practice

1 Fill in the blanks.

(a)

____ minutes

past ____

(b)

____ minutes

past ____

2 Write the time.

(a)

____

(b)

____

Complete Worksheet **1**: pages **125–126**

Telling and Writing Time

Lesson 2

Explore

Who is correct?

Master

1

The time is 50 minutes after 7 o'clock.
It is 50 minutes past 7.

Ruby is correct.

The time is 10 minutes before 8 o'clock.
It is 10 minutes to 8.

Oak is correct.

2

The time is 15 minutes before 5 o'clock.
It is 15 minutes to 5.

Guided Practice

1 Fill in the blanks.

(a)

____ minutes to ____

(b)

____ minutes to ____

2 Write the time.

(a)

____

(b)

____

Complete Worksheet **2**: pages **127–128**

Sequencing Events

Lesson 3

Explore

Place these events of Charles' day in order, starting from the earliest.

Master

1

Charles brushed his teeth at 7:45 in the morning.

He left for school at 8:30 in the morning.

He left school for home at 3:15 in the afternoon.

2  Charles brushed his teeth **before** he left for school.

 Charles left for school **after** he brushed his teeth.

Guided Practice

1 Below are three activities that Elliott did yesterday. Tell the time of each activity and fill in the blanks.

ate dinner

cycled

read a book

(a) The first activity Elliott did was ________.

(b) The next activity Elliott did was ________.

(c) The last activity Elliott did was ________.

2 The clocks show the different times the children finished their swimming lessons. Tell the time and fill in the blanks.

Lulu

Ruby

Jacob

(a) ______ finished their swimming lesson first.

(b) ______ finished their swimming lesson after Jacob.

(c) The last child to finish their swimming lesson was ______ .

Complete Worksheet **3**: pages **129–132**

Drawing Clock Hands

Lesson 4

Explore

Emma is setting the clock to show 3:45.
Where should she place the hour hand?

Should the clock look like this 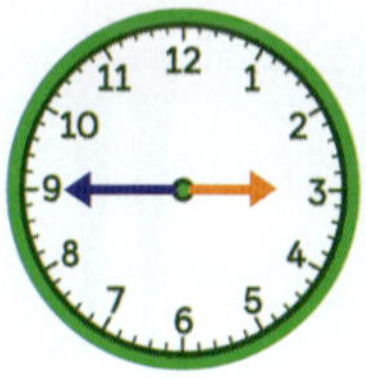 or like this  ?

Master

1 3:45 is close to 4:00.
The hour hand is closer to 4 than to 3.

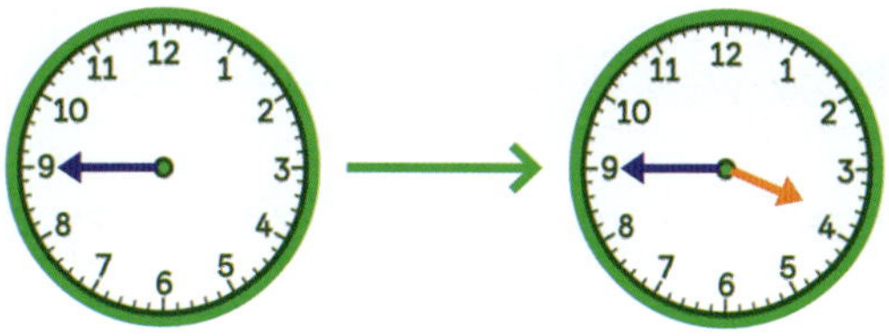

The clock should look like this.

The hour hand moves slowly from 3 to 4 during the hour that the minute hand moves after 3 o'clock.

2 Place the minute hand on the clock to show quarter past 1.

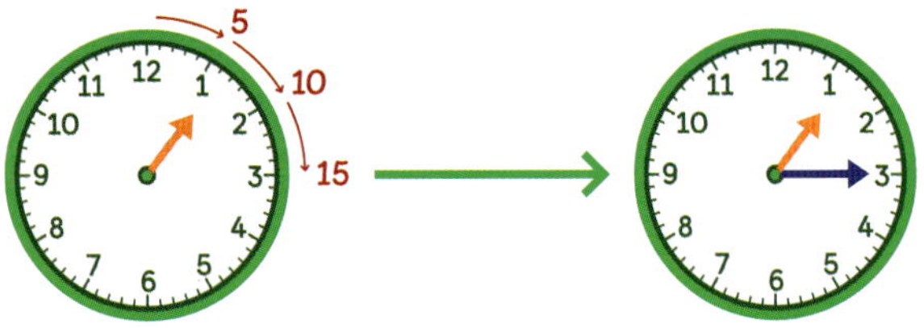

At quarter past 1, the minute hand is pointing at 3.

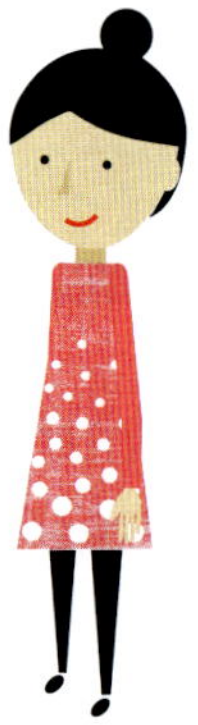

Activity Time

Partner work

You will need:

① Pick a  and say the time.

② Ask your partner to show you the time on the  and tell you an activity they do at that time.

For example:

7:30

③ Repeat the activity, taking turns.

Guided Practice

1 Which clock shows 4:55?

clock A

clock B

2 Draw the missing hour hand.

(a) half past 11

(b) 4:05

3 Draw the missing minute hand.

(a) 1:25

(b) 8:55

4 Draw the missing hour hand and minute hand.

(a) 3:15

(b) 7:20

Complete Worksheet **4**: pages **133–134**

Finding Durations of Time

Lesson 5

Explore

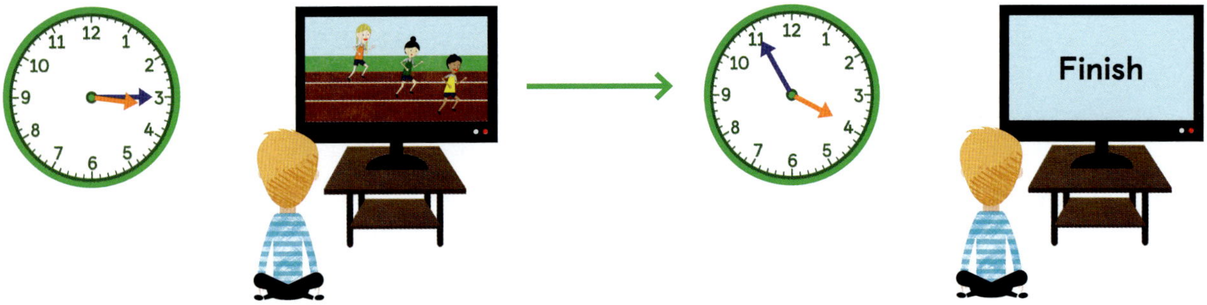

The TV programme started.

The TV programme finished.

How long was the TV programme?

Master

Method 1

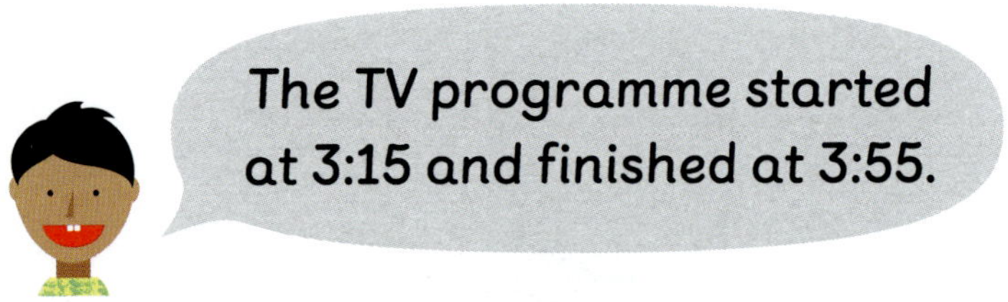

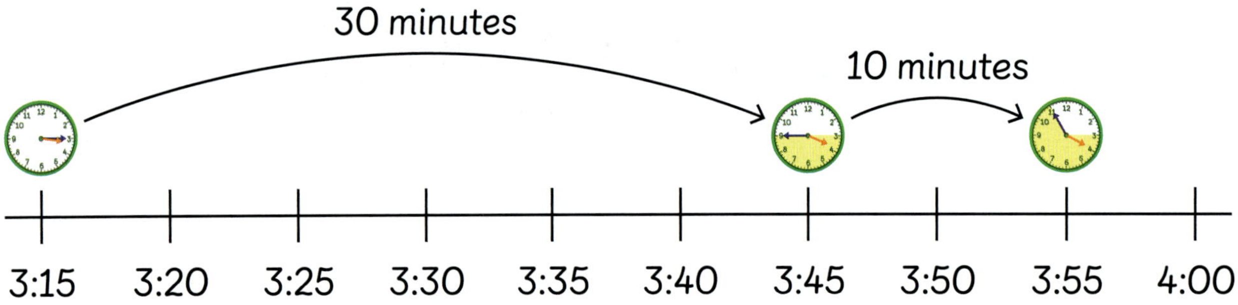

30 minutes + 10 minutes = 40 minutes

The TV programme was 40 minutes long.

Method 2

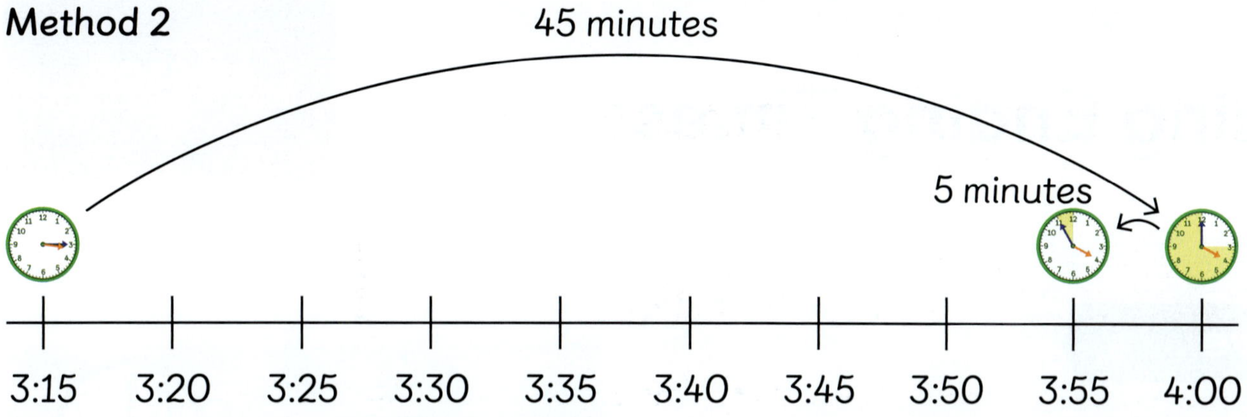

45 minutes − 5 minutes = 40 minutes

The TV programme was 40 minutes long.

Guided Practice

1 How long is each TV programme?

(a)

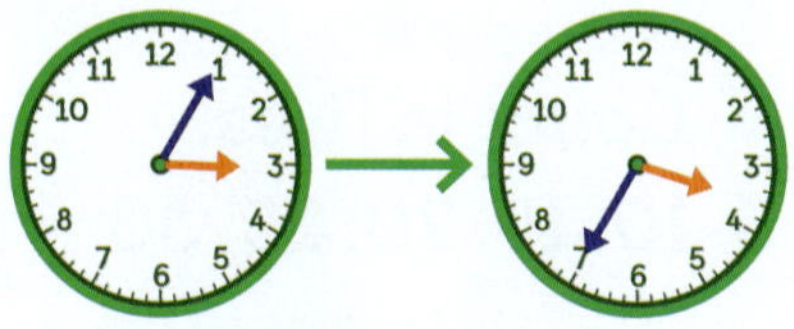

minutes

(b)

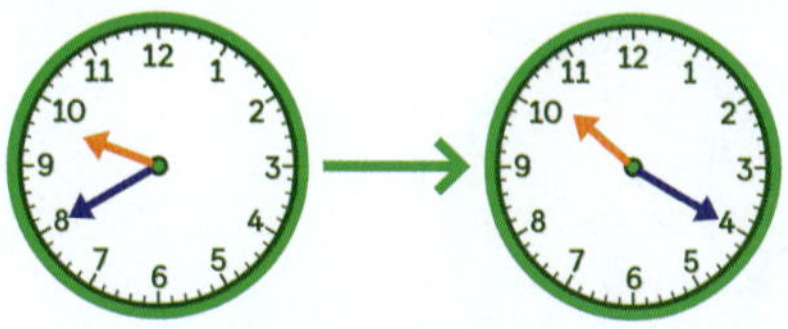

minutes

(c)

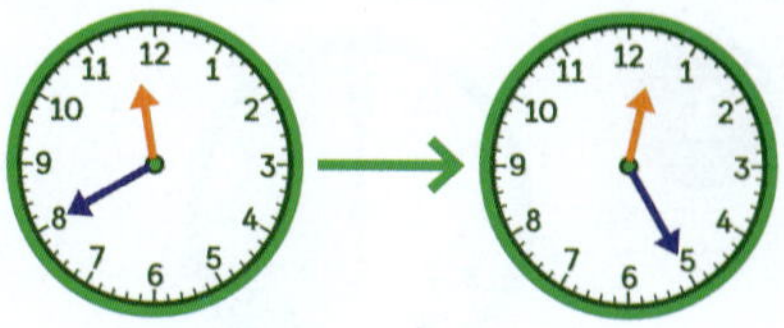

minutes

2 Hannah started her dance class at 3:45 in the afternoon.
The class ended at 4:30 in the afternoon.
Draw the hands on the clocks to show the start and end times.
Then, find the duration of the class.

class starts → class ends

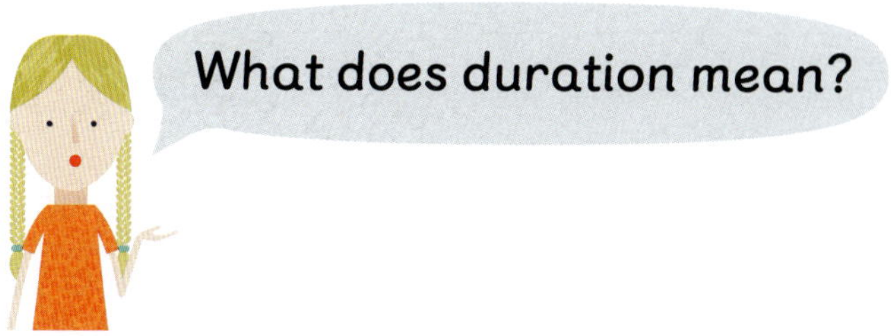

Complete Worksheet **5**: pages **135–136**

Finding Ending Times

Lesson 6

Explore

It takes Amira 30 minutes to run half of a race.
It takes her 1 hour to finish the race.
What time will it be when she is halfway through the race?
What time will she finish the race?

Master

1

The minute hand moves from 12 to 6 in 30 minutes.
The time will be 6:30 when Amira is halfway through the race.

2

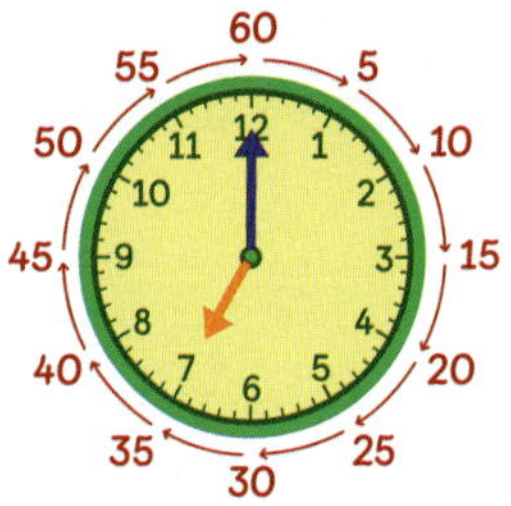

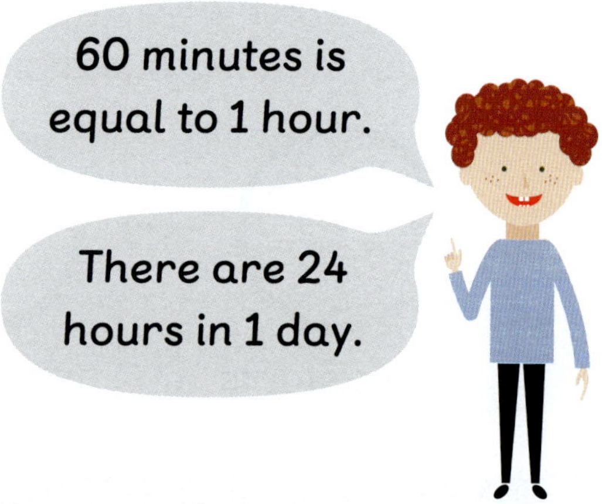

Amira finishes the race in 1 hour.
The hour hand moves from 6 to 7 in 1 hour.
The minute hand makes a complete round during the 60 minutes and returns to 12.
Amira will finish the race at 7 o'clock.

Guided Practice

1 

1 hour = ____ minutes

(a) What is the time 60 minutes after 7 o'clock? ____

(b) What is the time 1 hour after 7 o'clock? ____

(c) What do you notice about the two answers? ____

2 What will the time be in 30 minutes?
What will the time be in 1 hour?

(a) 

____
____

(b) 

____
____

3 What will the time be in 2 hours?
What will the time be in 4 hours?

(a) 

____
____

(b) 

____
____

Complete Worksheet **6**: pages **137–138**

Finding Ending Times

Lesson 7

Explore

Emma, Charles and Jacob get on the train.
Emma's journey is 3 hours.
Charles' journey is 30 minutes.
Jacob's journey is 55 minutes.
At what time do they reach their destinations?

Master

1

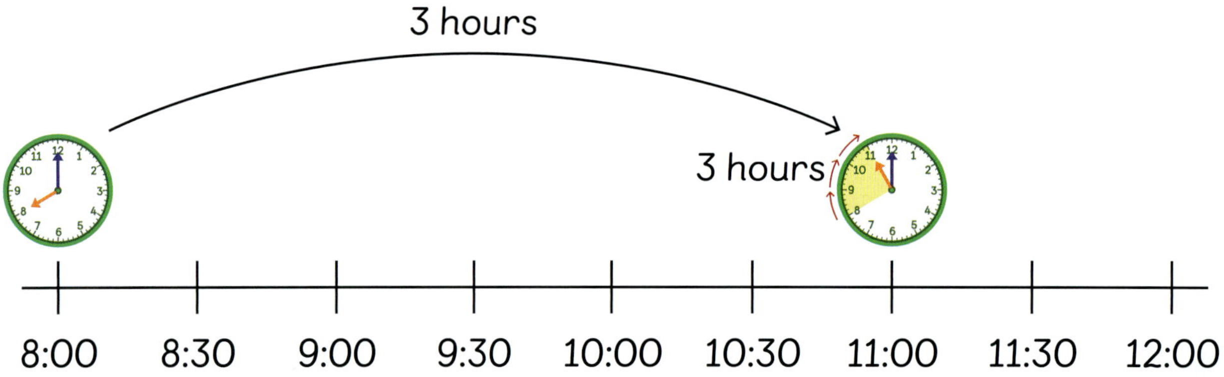

3 hours after 8:00 is 11:00.
Emma arrives at her destination at 11:00.

2

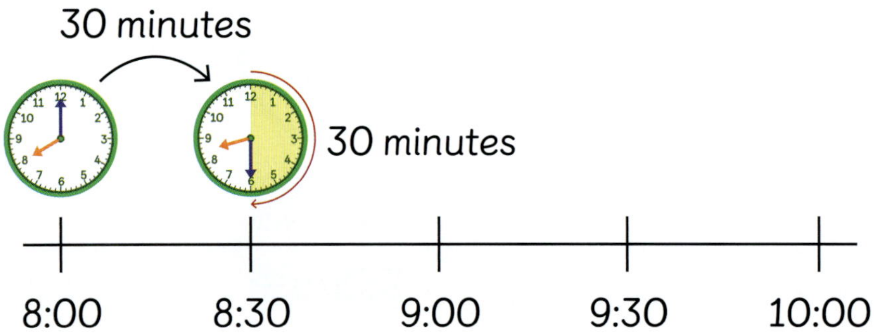

30 minutes after 8:00 is 8:30.
Charles arrives at his destination at 8:30.

3

55 minutes after 8:00 is 8:55.
Jacob arrives at his destination at 8:55.

Guided Practice

At what time did each child finish reading their storybooks?
Show the end times on the clocks below.

1

Ruby finished reading her storybook at ______.

2

Sam finished reading his storybook at ______.

3

Lulu finished reading her storybook at ______.

Complete Worksheet **7**: pages **139–140**

Lesson 8

Finding Starting Times

Explore

Oak, Elliott and Amira took a break at 11:00 from working on their school projects.

At what time did each child start working on their project?

Master

1 At what time did Oak start working on her project?

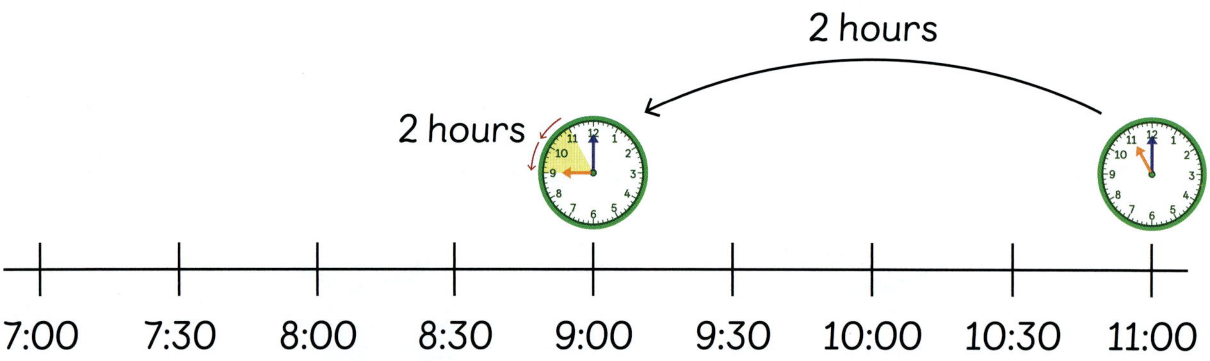

2 hours before 11:00 is 9:00.

Oak started working on her project at 9:00.

2 At what time did Elliott start working on his project?

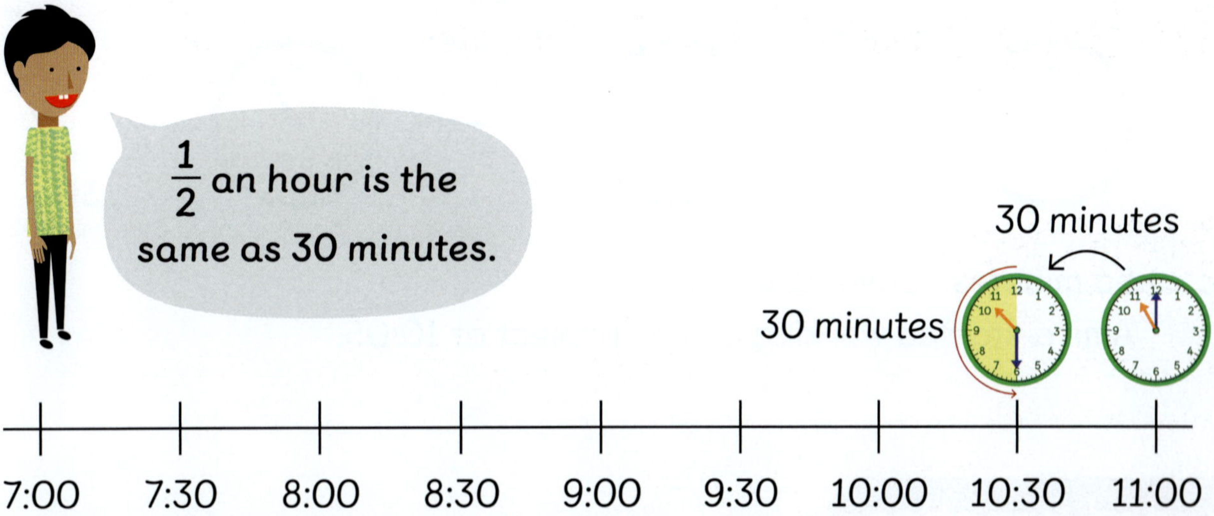

30 minutes before 11:00 is 10:30.
Elliott started working on his project at 10:30.

3 At what time did Amira start working on her project?

Method 1

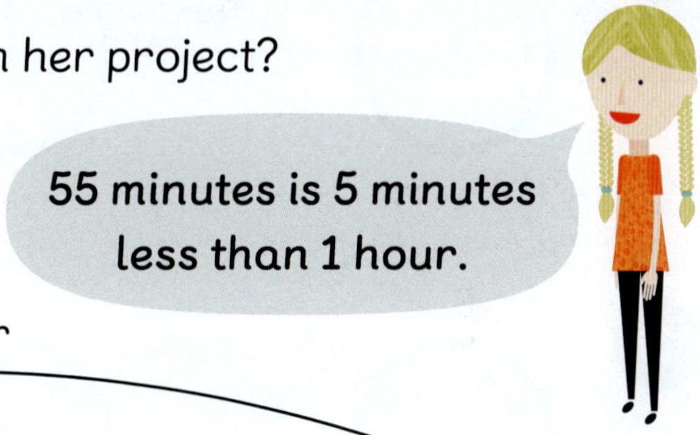

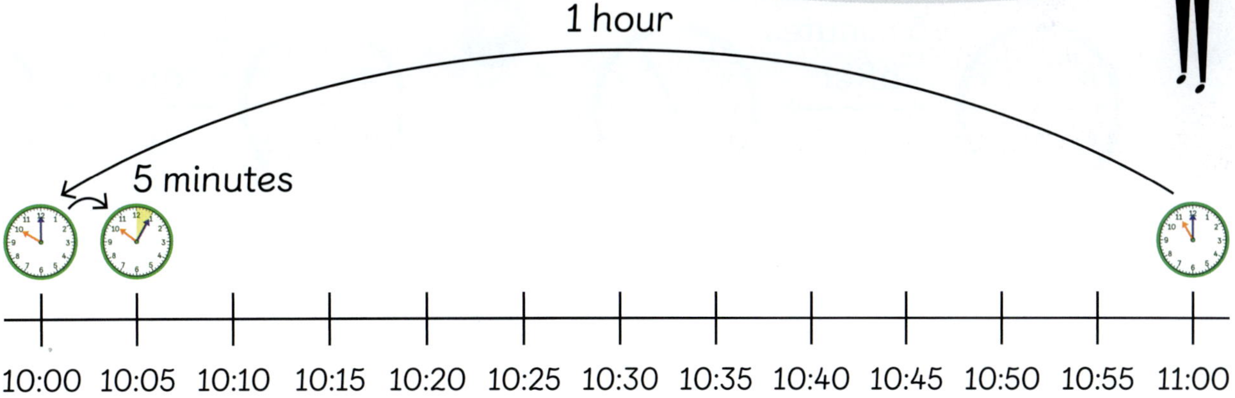

55 minutes before 11:00 is 10:05.

Method 2

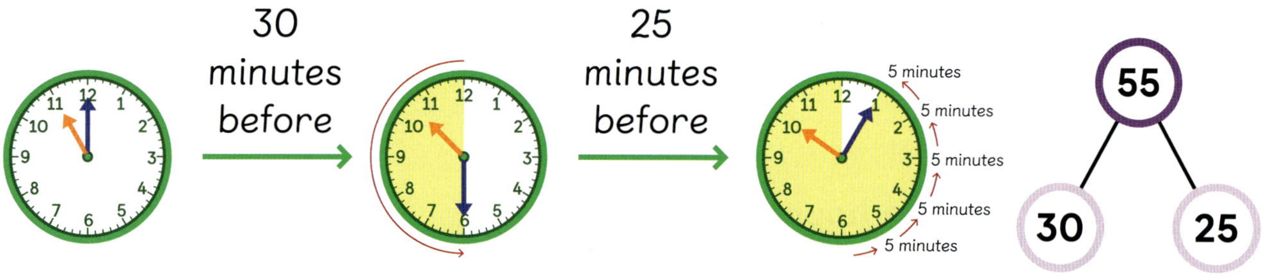

55 minutes before 11:00 is 10:05.
Amira started working on her project at 10:05.

Guided Practice

Draw the clock hands to show the start times.

1

2

3

4

5

6

Complete Worksheet **8**: pages **141–142**

Comparing Durations of Time

Lesson 9

Explore

Sam

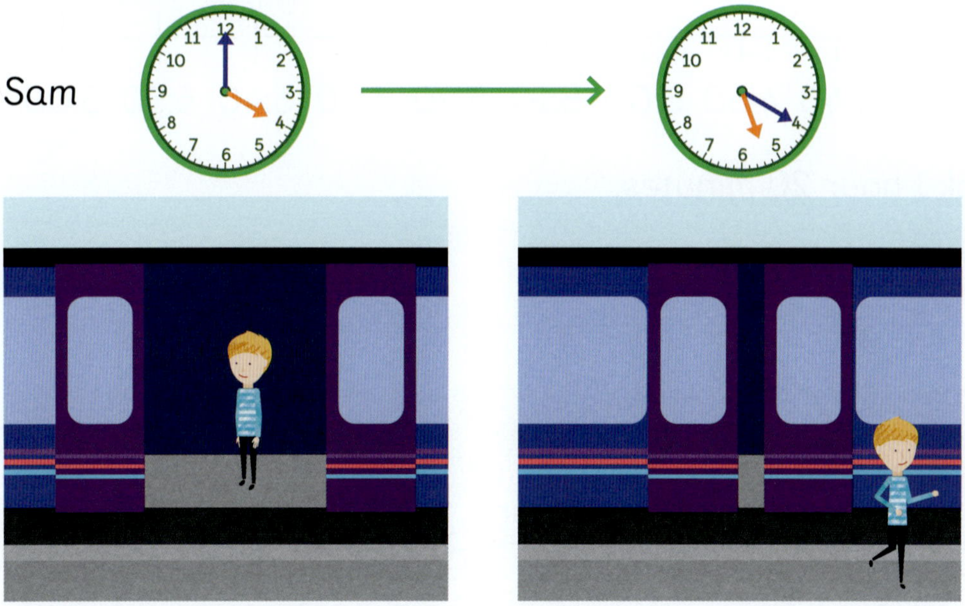

Lulu

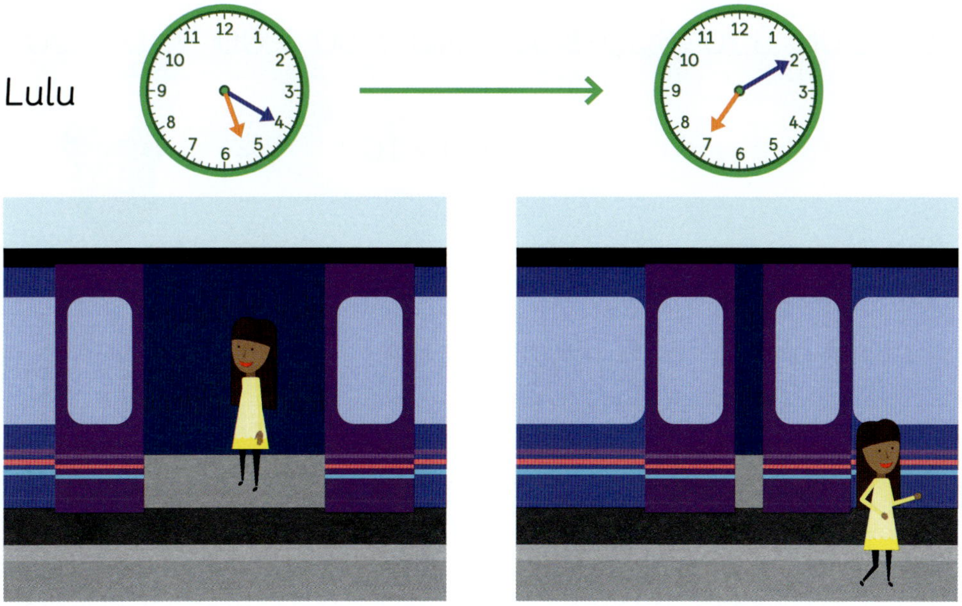

Whose train journey took a greater amount of time?

Master

1

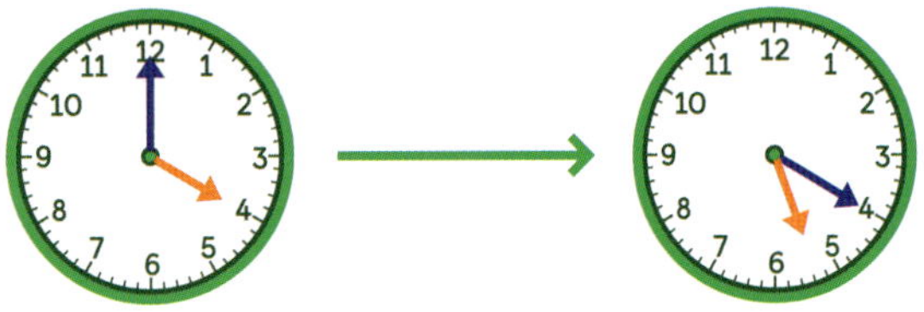

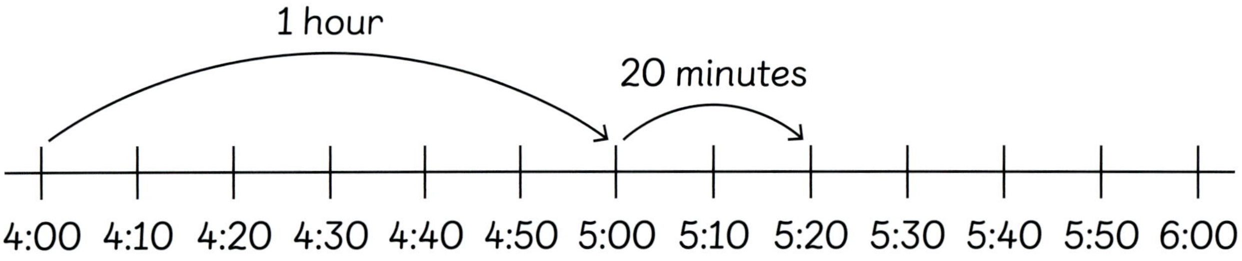

Sam's journey took 1 hour 20 minutes.

2

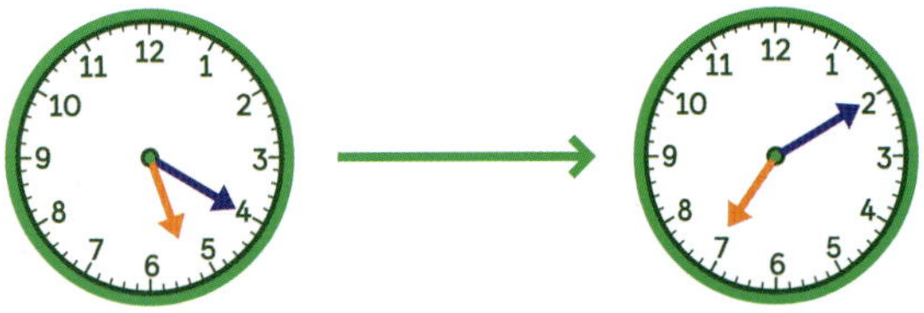

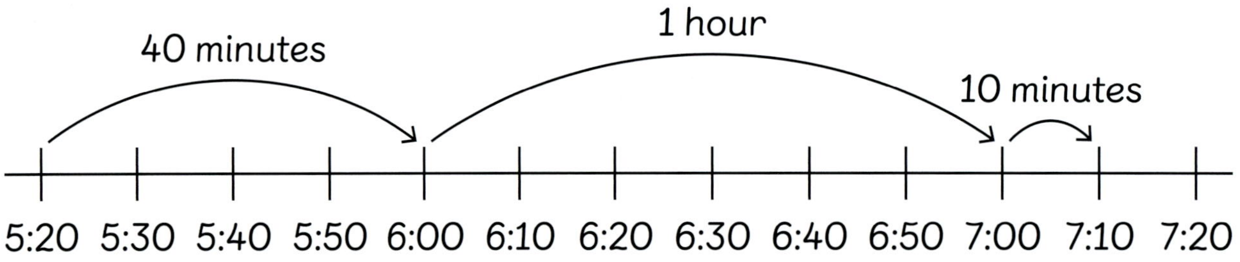

Lulu's journey took 1 hour ____ minutes.

____ 's journey took a greater amount of time.

Guided Practice

1 The clocks show the start and end times of the children's bus journeys. Whose bus journey took longer?

(a) start — end

Jacob — Emma

__________'s bus journey took longer.

(b) start — end

Amira — Ruby

__________'s bus journey took longer.

2 The clocks show the start and end times of a wall climbing activity completed by three children.

Elliott

Emma

Lulu

(a) Who spent the most amount of time climbing the wall?

(b) Who spent the least amount of time climbing the wall?

Complete Worksheet **9**: pages **143–144**

Mind Challenge

Ravi, Elliott and Holly looked at the clock when they finished their homework.

Ravi

Elliott

Holly

Ravi started his homework three hours after he finished lunch at 2 o'clock.
Elliott started his homework one hour after Ravi.
Holly started her homework half an hour before Ravi.
How much time did each child spend doing their homework?

My Maths Journal

Time yourself doing these tasks and record the start times, end times and durations in your journal.

- Read five pages of your reading book.
- Sing your favourite song.
- Write a short story on a topic of your choice.

Review

I can...

- [ ] tell and write the time to 5 minutes.
- [ ] use quarter past and quarter to the hour when telling time.
- [ ] compare and sequence intervals of time.
- [ ] draw the hands on a clock face to show time to 5 minutes.
- [ ] find durations of time.
- [ ] find the end and start times of activities.
- [ ] tell the number of minutes in an hour.
- [ ] tell the number of hours in a day.

How much water did Sam use to wash the van?

Chapter 15
Volume

Lesson 1

Comparing Volume

Explore

Jacob pours some juice into different glasses for his friends.

Sam Ruby Lulu

Who has the greatest volume of juice?

Master

1 To compare the volume of juice in each glass, the juice is poured into identical beakers.

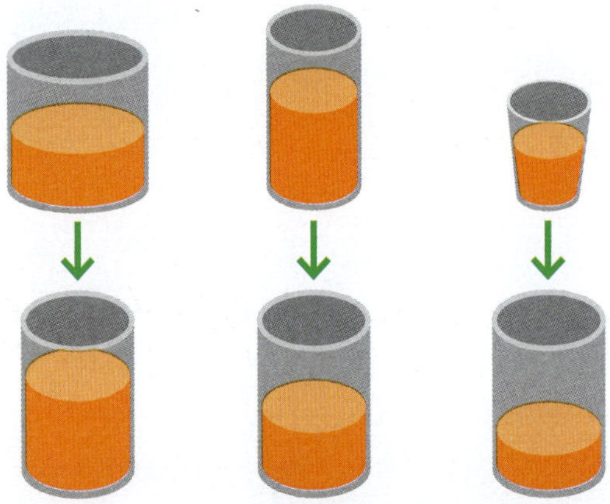

Sam Ruby Lulu

Sam has the **greatest** volume of juice.
Lulu has the **least** volume of juice.

2 The volume of juice from Sam's glass is **greater than** the volume of juice from Ruby's glass.

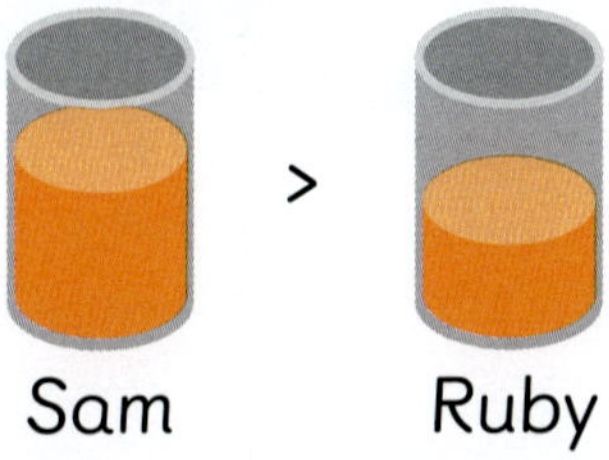

The volume of juice from Lulu's glass is **less than** the volume of juice from Ruby's glass.

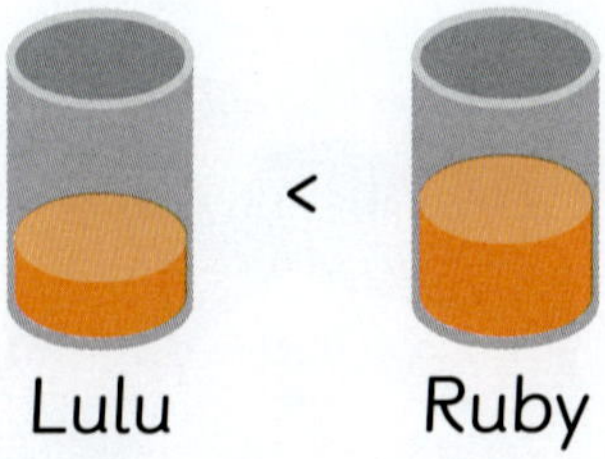

Guided Practice

1 Compare the volume of water in the containers using >, < or =.

(a) 

(b) 

2 Fill in the blanks using A, B or C.

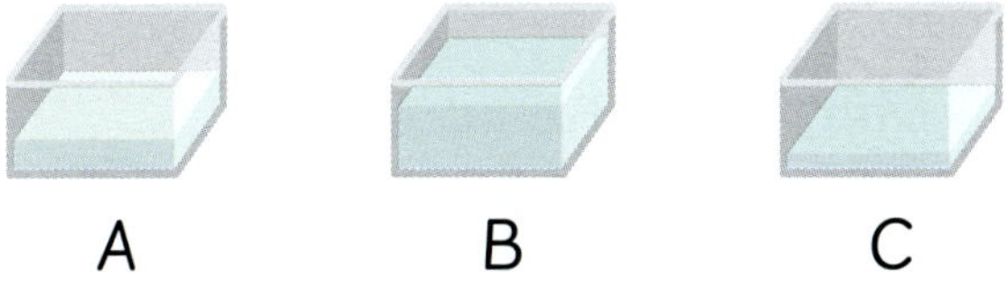

(a) Container ______ has the greatest volume of water.

(b) Container ______ has the least volume of water.

(c) Container A has less water than container ______.

Complete Worksheet 1: pages 149–150

Lesson 2

Comparing Volume

Explore

How can Holly compare the volume of water in each vase using recyclable plastic cups?

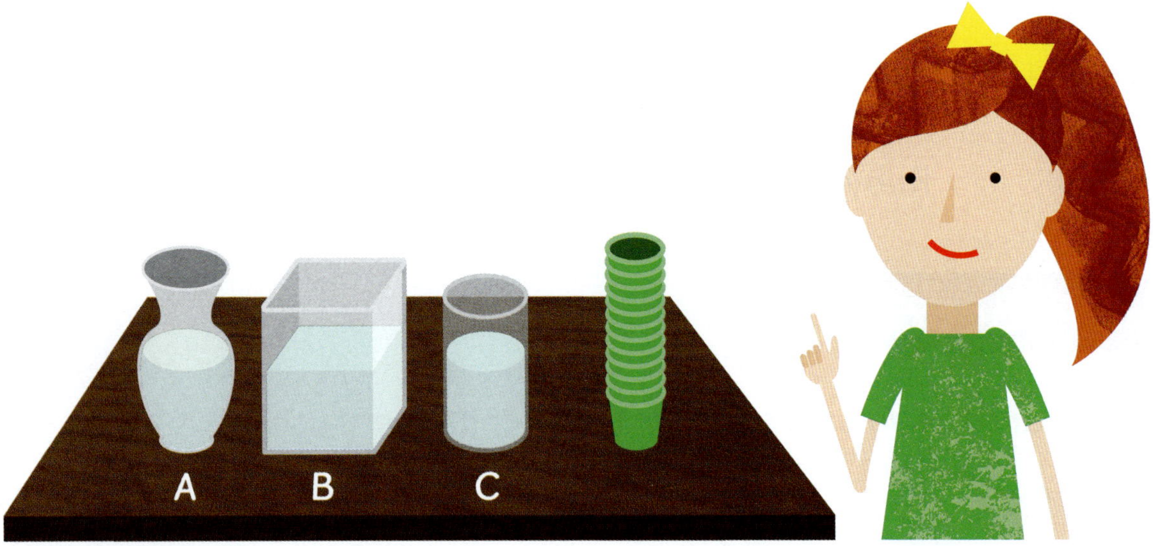

Master

Use  as 1 unit.

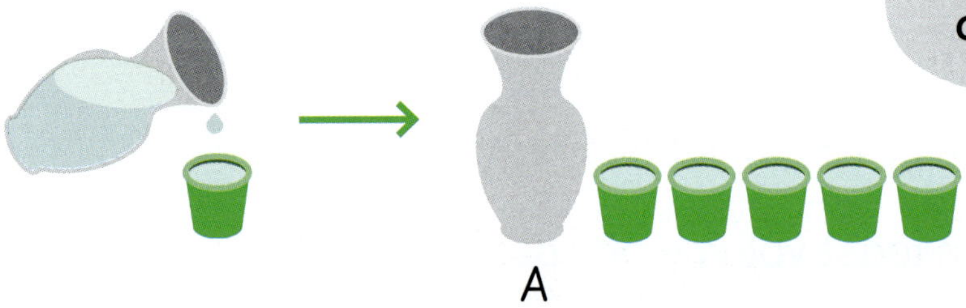

The water in vase A fills 5 cups.
The volume of water in vase A is 5 units.

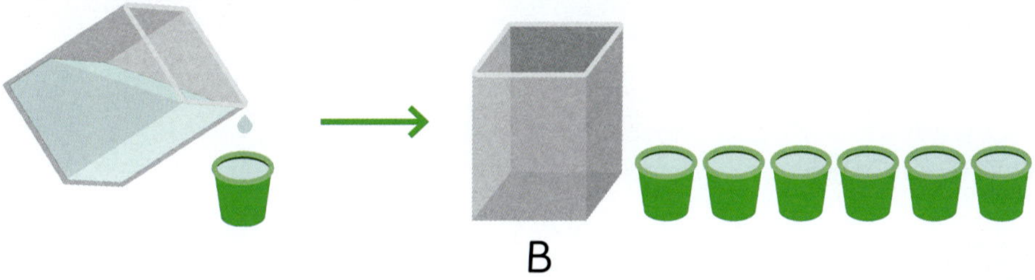

The volume of water in vase B is 6 units.

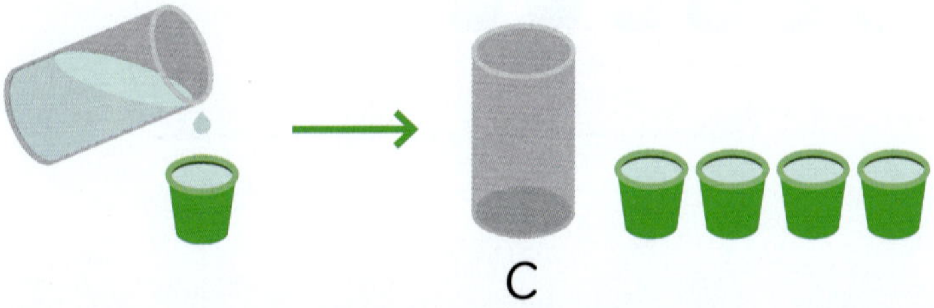

The volume of water in vase C is 4 units.

Vase B has the greatest volume of water.
Vase C has the least volume of water.

Activity Time

Partner work

You will need:

1. Choose 3 different containers.
2. Fill the containers with water to different levels.
3. Ask your partner to guess which container has the greatest volume of water.
4. Use the cups to measure the volume of water in each container.

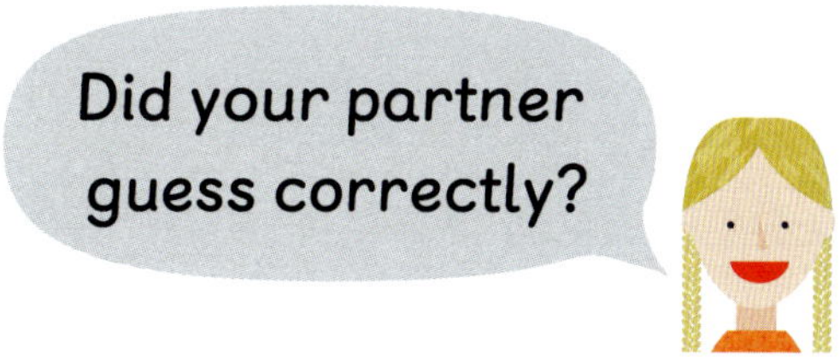

5. Repeat the activity, taking turns.

Guided Practice

1 Use A, B or C to fill in the blanks.

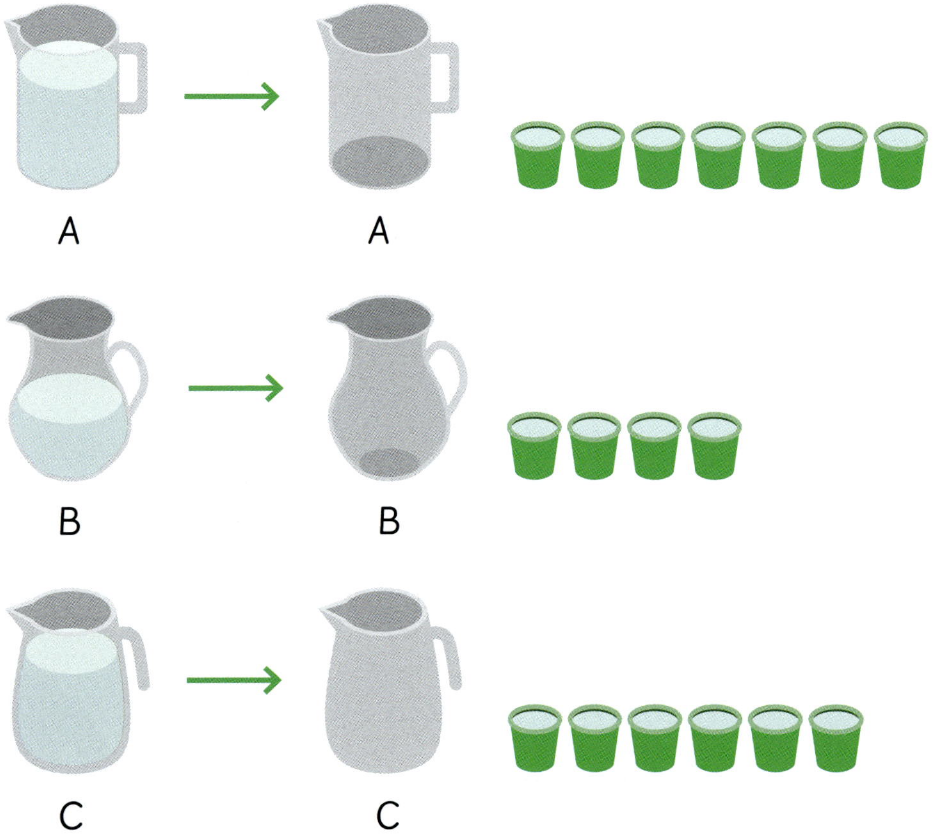

(a) Jug ____ has the greatest volume of water.

(b) Jug ____ has the least volume of water.

(c) The volume of water in jug ____ is 6 units.

2 Use >, < or = to compare the volume of water in the jugs.

(a)

(b)

Complete Worksheet 2: pages 151–152

Measuring Volume in Litres

Lesson 3

Explore

Amira has 1 litre of apple juice.

Which bottles contain more or less than 1 litre of juice?

Master

1 Amira pours the apple juice into a 1-litre beaker.

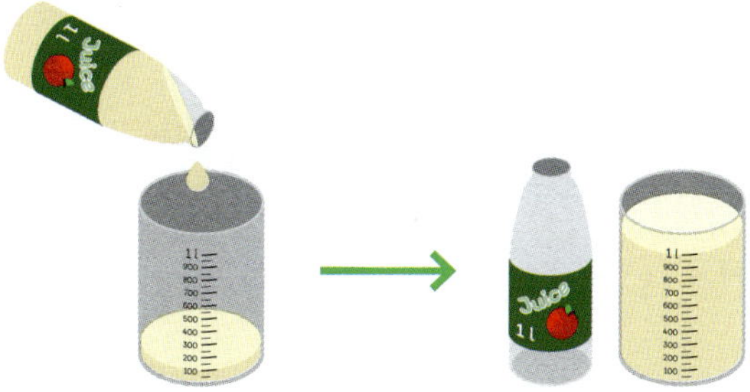

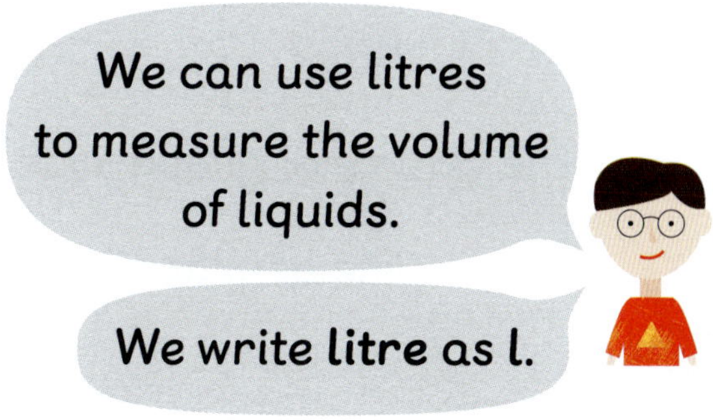

There is 1 l of apple juice in the bottle.

2 Measure the volume of lemon juice in the bottle.

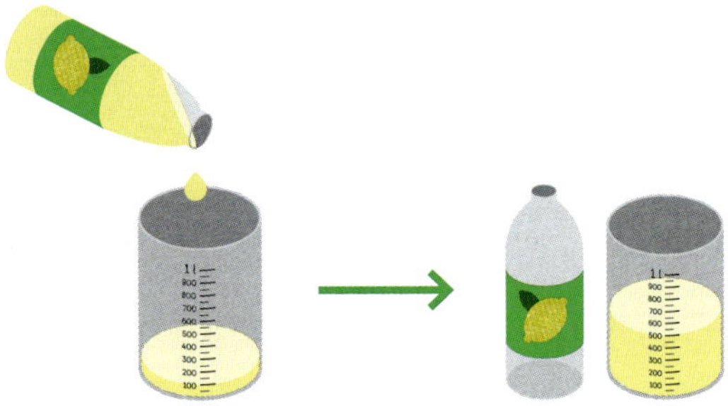

There is **less than** 1 l of lemon juice in the bottle.

3 Measure the volume of orange juice in the bottle.

There is **more than** 1 l of orange juice in the bottle.

4 Measure the volume of grape juice in the bottle.

The volume of grape juice in the bottle is ______ than 1 l.

The volume of grape juice in the bottle is ______ than 2 l.

Guided Practice

1 Use **more** or **less** to fill in the blanks.

(a)

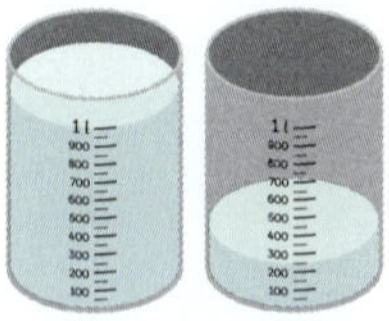

The beakers contain ______ than 1 l of water.

(b)

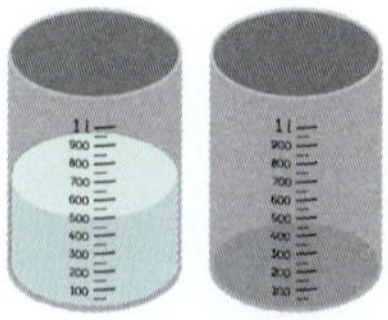

The beakers contain ______ than 1 l of water.

(c)

The volume of pineapple juice in the bottle is ______ than 1 l.

2 The water in the bottles was poured into 1-l beakers.
Find the volume of water that was in each bottle.

A

B

Bottle A had ______ l of water.

Bottle B had ______ l of water.

Complete Worksheet **3**: pages **153–154**

Measuring Volume in Millilitres

Lesson 4

Explore

Miss Fathima measures 3 different bottles of food colouring during a maths lesson.
She empties all the red food colouring from a bottle into a beaker.
What is the volume of red food colouring in the bottle?

Master

1 Find the volume of the red food colouring.

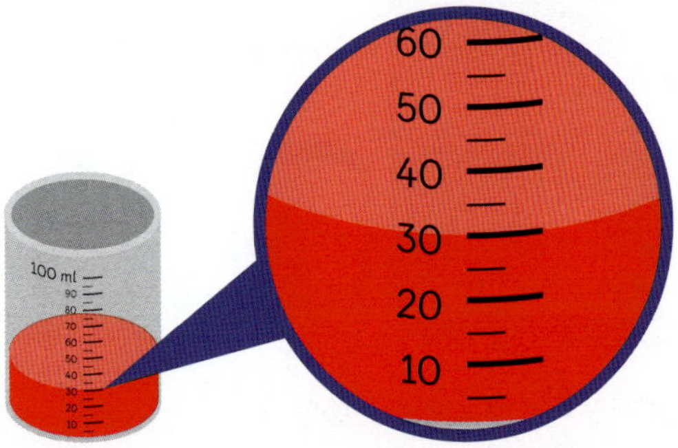

The volume of red food colouring in the bottle is 30 ml.

2 Find the volume of orange food colouring.

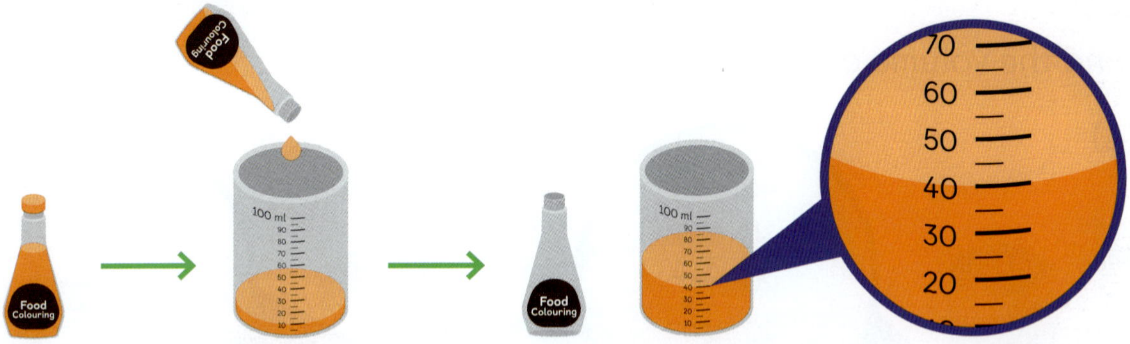

There is [ ] ml of orange food colouring in the bottle.

3 The bottle of green food colouring is full.
Empty the full bottle into the beaker.
What is the capacity of the bottle?

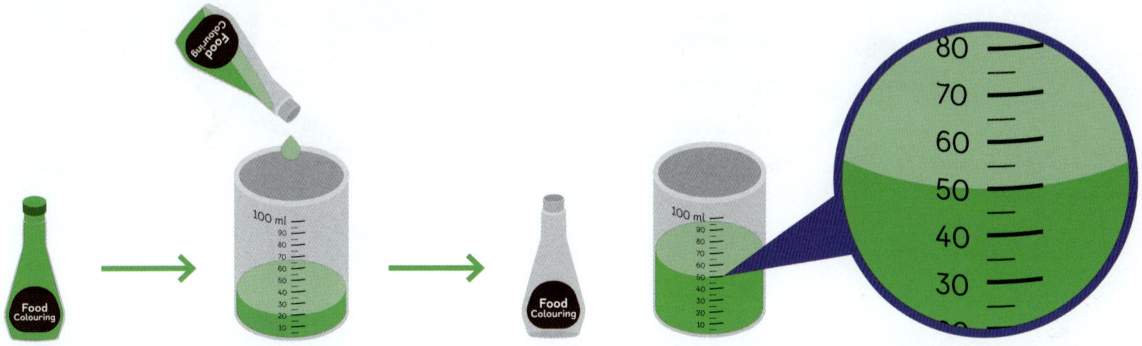

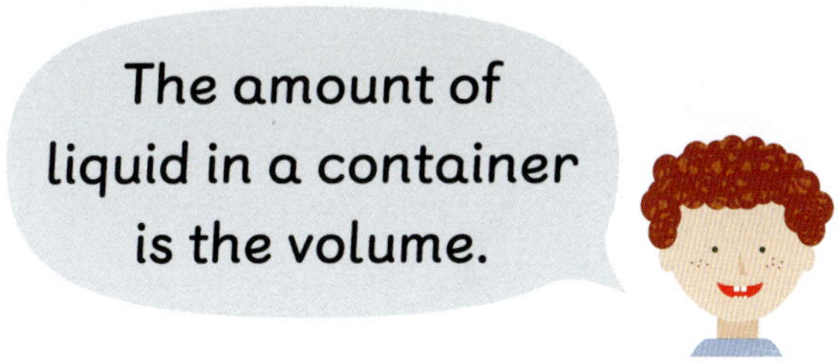

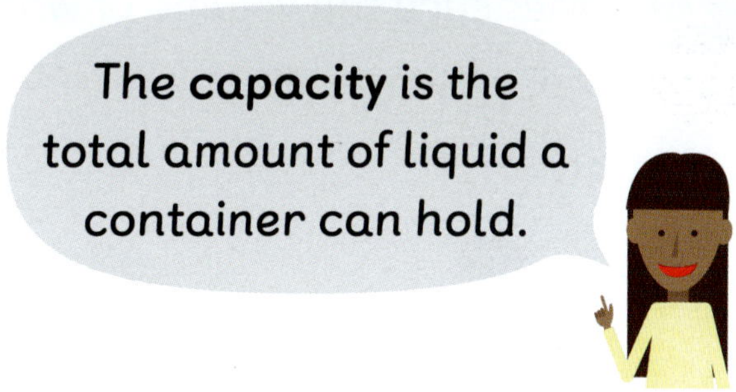

The capacity of the bottle is 50 ml.

4 Compare the volume of the food colourings using > or <.

Guided Practice

1 Find the volume of water in each beaker.

(a)

ml

(b)

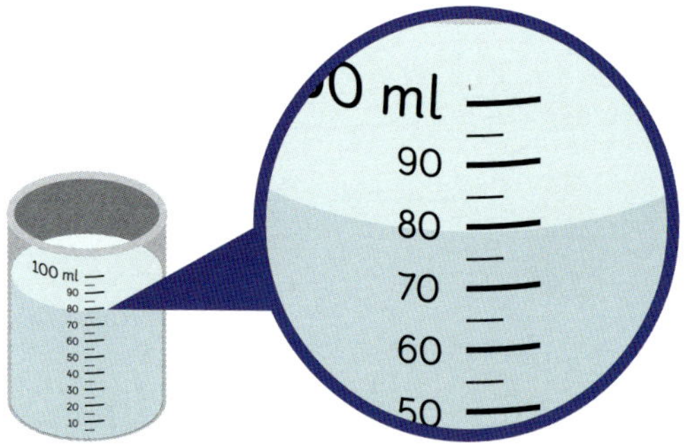

ml

(c)

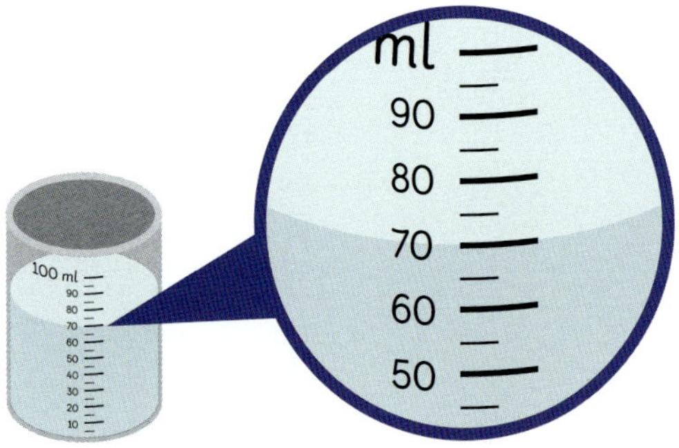

ml

(d)

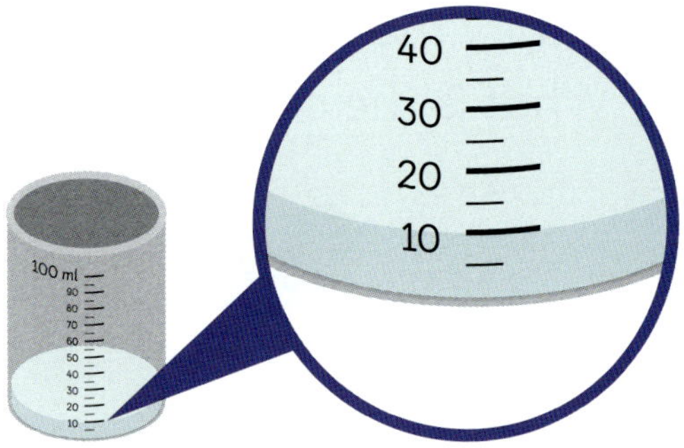

ml

2 Measure the volume of water in the bottle.

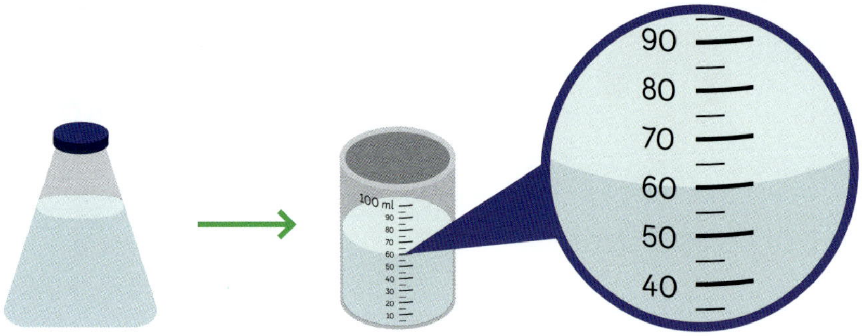

volume of water = ml

3 Measure the capacity of the bottle.

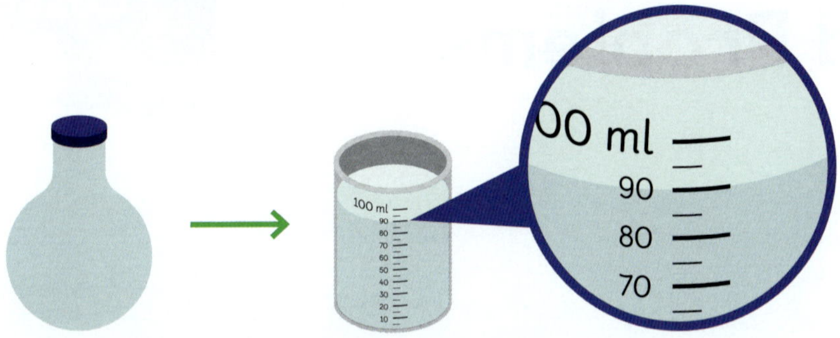

capacity of the bottle = ☐ ml

4 Compare the volumes using > or <.

(a)

80 ml ☐ 90 ml

(b)

60 ml ☐ 40 ml

Complete Worksheet **4**: pages **155–156**

Lesson 5

Solving Word Problems

Explore

Ruby and Ravi have both found very small containers. They want to know how much water each of their containers can hold. They each completely fill their containers with water and pour the water into the beakers.
What is the total capacity of both their containers?

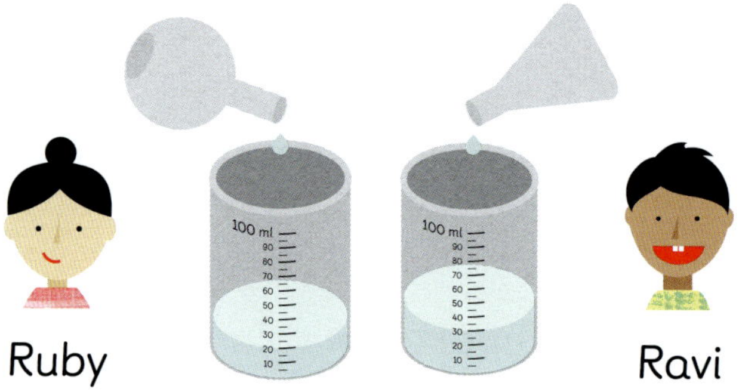

Master

1

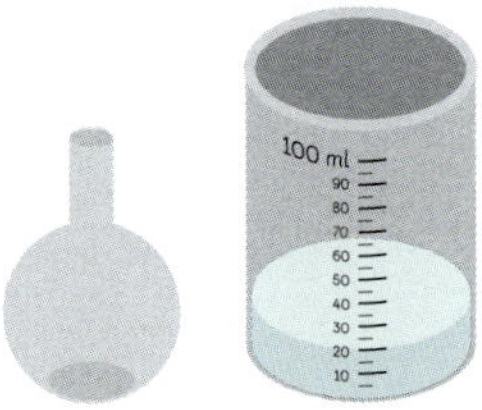

The capacity of Ruby's container is 20 ml.

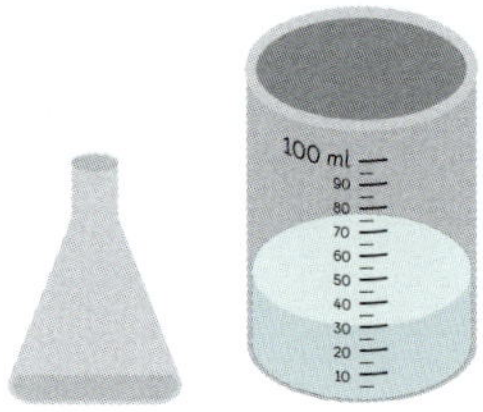

The capacity of Ravi's container is 30 ml.

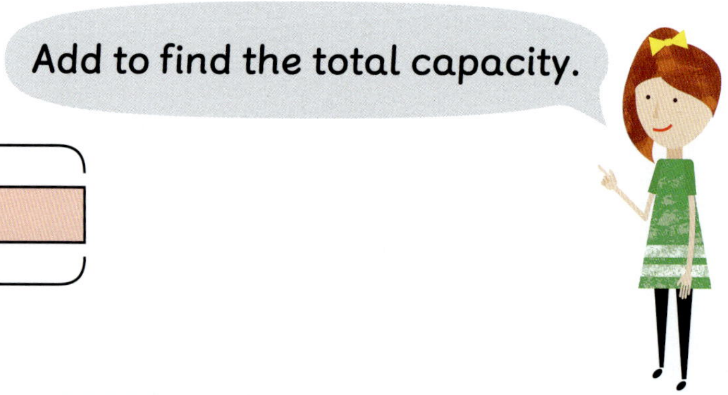

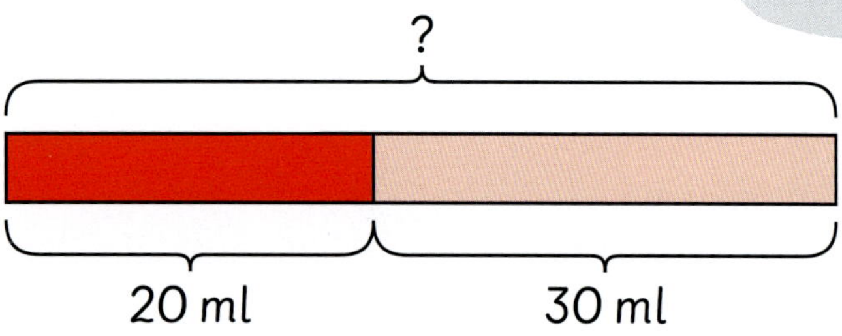

20 + 30 = 50
The total capacity of both containers is 50 ml.

2 A water dispenser in an office contains 20 l of water.
At the end of the day, 4 l are left.
How much water did the office workers drink?

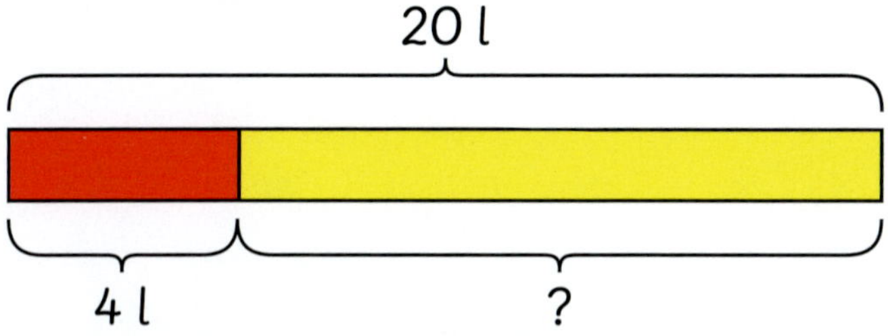

☐ ○ ☐ = ☐

The office workers drank ☐ l of water.

Guided Practice

35 l of rainwater has been collected in a container in a school's garden.
The teachers use 16 l in the morning to water the plants.
They use the remaining water in the afternoon.
What is the volume of water the teachers use in the afternoon?

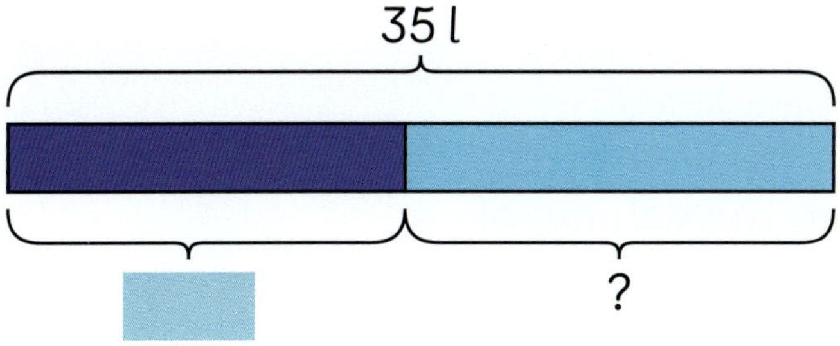

☐ ○ ☐ = ☐

The teachers use ☐ l of water in the afternoon.

Complete Worksheet **5**: pages **157–158**

Solving Word Problems

Lesson 6

Explore

A painter used 21 l of paint to paint the first floor of a hotel. He used 3 l more paint to paint the second floor of the hotel.

What volume of paint did he use altogether?

Master

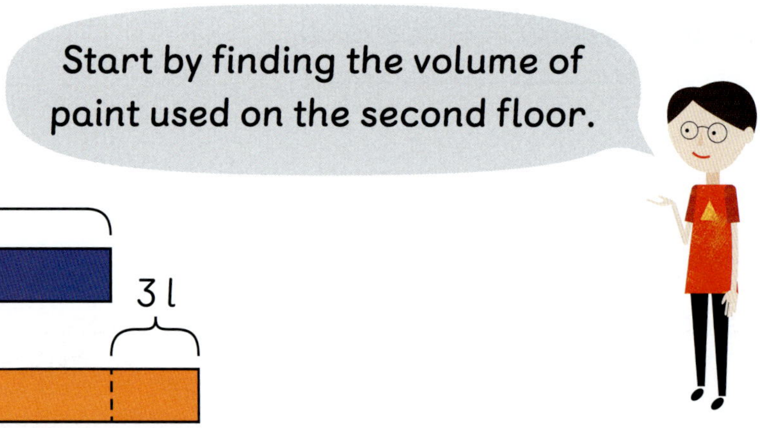

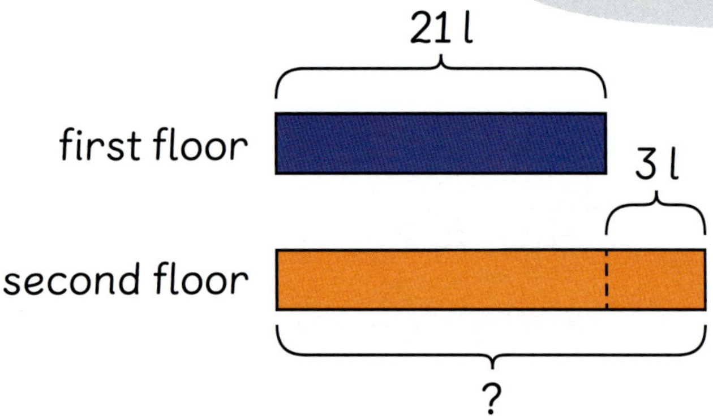

21 + 3 = 24

The painter used 24 l of paint to paint the second floor.

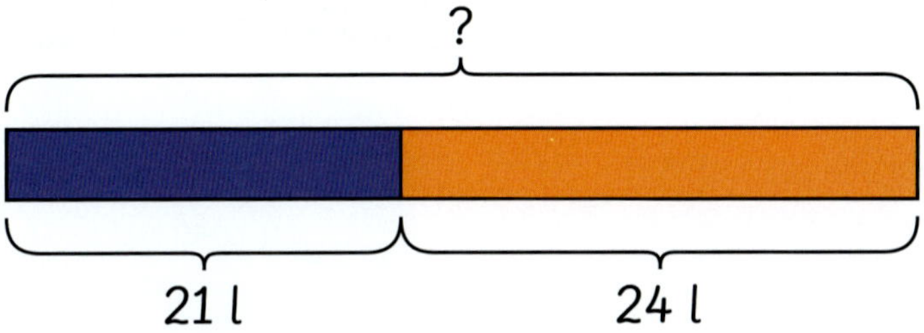

21 + 24 = 45

He used 45 l of paint altogether.

Guided Practice

There is 17 l of water in Lulu's fish tank.
Her fish tank has 4 l less water than Sam's fish tank.
What is the total volume of water in both fish tanks?

Complete Worksheet **6**: pages **159–160**

Solving Word Problems

Lesson 7

Explore

Mr Nightingale fills 4 large containers with water for a camping trip.
Each container has 10 l of water in it.

What is the total volume of water Mr Nightingale takes on the camping trip?

Master

1 **Method 1** Add.

10 + 10 + 10 + 10 = 40

Method 2 Multiply.

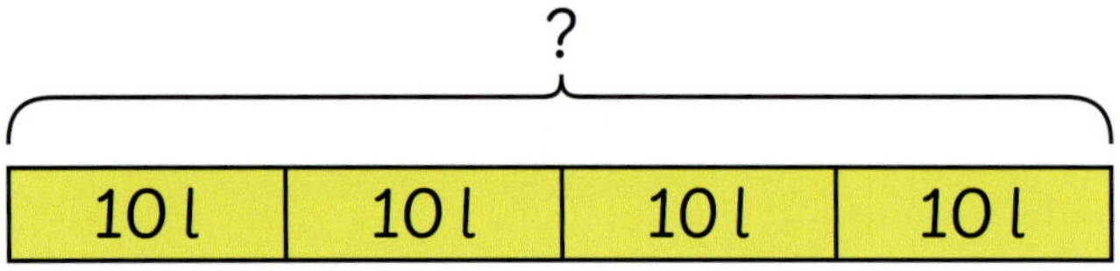

4 × 10 = 40

Mr Nightingale takes 40 l of water on the camping trip.

2 Amira's mum used 12 l of paint to paint 6 walls in her house.
She used an equal amount of paint on each wall.
How many litres of paint did she use to paint each wall?

☐ ◯ ☐ = ☐

Amira's mum used ☐ l of paint to paint each wall.

Guided Practice

A mechanic buys 6 containers of car oil.
Each container has 8 l of car oil.

1 How many litres of car oil does the mechanic have altogether?

2 The mechanic uses all of the oil equally between 12 cars.
How many litres of oil does he use on each car?

Complete Worksheet **7**: pages **161–162**

Mind Challenge

The vase contains 2 l less water than the bowl.
The total volume of water in the vase and the bowl is 8 l.
The watering can contains twice as much water as the bowl.

What is the volume of water in each container?

My Maths Journal

Jacob uses several buckets of water to wash his mum's car. Would you measure the volume of water Jacob uses in litres or millilitres?
Explain your answer.

I can...

Review

- ☐ compare volume.
- ☐ measure volume in millilitres (ml) and litres (l).
- ☐ measure the capacity of a container.
- ☐ solve word problems on volume.